新工科建设之路·计算机类专业系列教材

微服务分布式架构基础与实战
——基于 Spring Boot + Spring Cloud

张方兴 编著

电子工业出版社
Publishing House of Electronics Industry
北京·BEIJING

内 容 简 介

本书以分布式架构结合微服务实例的方式，介绍 Spring Boot + Spring Cloud 的基础知识、架构顺序和操作方法。通过学习前 5 章，可基本搭建 Consul 集群、多个微服务、微服务间通信、负载均衡、断路器的分布式基本结构；后 6 章主要介绍如何编写微服务业务代码，包括 Spring Boot、MySQL、Redis、缓存一致性、事务、异步线程池、分布式消息通信、分布式任务调度管理及 FastDFS 分布式文件管理；第 12 章对微服务分布式架构进行了扩展与总结。

本书可作为高等学校计算机、软件工程等专业 Java 架构相关课程的教材，也可供对微服务分布式架构感兴趣的人员参考阅读。

未经许可，不得以任何方式复制或抄袭本书之部分或全部内容。
版权所有，侵权必究。

图书在版编目(CIP)数据

微服务分布式架构基础与实战：基于 Spring Boot+Spring Cloud/张方兴编著. —北京：电子工业出版社，2020.3
ISBN 978-7-121-38413-4

I. ①微… II. ①张… III. ①互联网络—网络服务器 IV. ①TP368.5

中国版本图书馆 CIP 数据核字(2020)第 022234 号

责任编辑：章海涛
文字编辑：张　鑫
印　　刷：涿州市般润文化传播有限公司
装　　订：涿州市般润文化传播有限公司
出版发行：电子工业出版社
　　　　　北京市海淀区万寿路 173 信箱　　邮编：100036
开　　本：787×1092　1/16　印张：17　字数：436 千字
版　　次：2020 年 3 月第 1 版
印　　次：2025 年 1 月第 8 次印刷
定　　价：59.00 元

凡所购买电子工业出版社图书有缺损问题，请向购买书店调换。若书店售缺，请与本社发行部联系，联系及邮购电话：(010)88254888，88258888。
质量投诉请发邮件至 zlts@phei.com.cn，盗版侵权举报请发邮件至 dbqq@phei.com.cn。
本书咨询联系方式：192910558(QQ 群)。

前言

目前，Spring Boot + Spring Cloud 架构已经成为 Java 程序员的必备技能之一，刚开始学习时看到琳琅满目的 Spring 全家桶，可能会感到无从下手。如果只了解微服务中的各知识点，而忽略了以微服务分布式架构的方式学习系统的架构顺序，初学者可能就不知道如何使用微服务构建分布式系统。为了使读者更快掌握 Spring Boot + Spring Cloud 的基础知识与架构方法，本书的章节顺序即应用系统的架构顺序。

分布式可以理解为人体器官，人体可看成分布式系统，大脑是注册中心的集群，四肢与器官是提供服务的微服务，前进的距离是微服务运行之后的返回值，消耗的体力是微服务中处理的逻辑，影响的记忆是某些微服务对数据库的增删改查。分布式可看成一种思想，而 Spring Cloud 与 Spring Boot 是实现了这种思想的工具。

无论多复杂的分布式应用程序，整合多少个服务器，调用多少种服务接口，使用多少种协议，使用集群还是高可用的何种架构方式，使用客户端还是服务端的何种负载均衡，使用哪个消息中间件、哪个数据库集群，使用搜索引擎/非关系型数据库/时序性数据库/关系型数据库/文件管理等多少种存储介质，其分布式本质都是分布式自身的思想。建议读者动手将本书实例都敲在 IDE 中，在 Linux 服务器上搭建各集群，那么上面那些看起来颇具难度的问题，都将不会是难题。

本书主要内容如下。

介绍微服务分布式的相关概念，搭建第一个微服务项目，了解微服务项目的运行过程。通过微服务整合 Consul 注册中心，搭建第一个微服务+注册中心的分布式系统。多个微服务与 Consul 注册中心相连，彼此通信，微服务获得彼此的接口及地址，调用彼此的接口与服务。然后无可避免地需要处理彼此通信时报错的情况，以免单一微服务无法正常提供服务，导致整个分布式系统的瘫痪。此时采用 Ribbon 客户端负载均衡方案，依靠多台服务器部署多个相同微服务项目，以提高系统的性能。在分布式通信不足以解决全部报错问题时，可选择 Hystrix 进行更精细划分，保证在任何一台微服务出现问题时，系统整体仍然能正常运行。

至此初步搭建了分布式系统，然后开始增加微服务的增删改查等业务功能。在系统处理增删改查过程的同时，还需要事务功能的支撑与管理。在初步增删改查后，依靠微服务的缓存增加微服务的性能，同时需要确保 Redis 与 MySQL 之间的增删改查一致性。另外，如果有特殊业务需求，可由分布式消息通信彼此协作、沟通、处理。

在处理基本业务后，还会有定时任务等业务需求，此时需要微服务的任务调度进行处理。而单节点任务调度的微服务可能会有宕机或重复执行等相关问题，只有使多台微服务同时进行任务调度，且彼此协同的情况下，才能解决此类问题。这时需要 Quartz 分布式任务调度解决多个任务调度间彼此协同、相互管理等问题。若有文件上传、下载等相关需求，需要使用微服务的文件上传管理。单节点文件上传可能存在磁盘空间不足或不易管理等问题，需要 FastDFS 分布式文件管理以解决多台文件管理服务器彼此协同、磁盘扩容等问题。

在对整体分布式微服务编程后，需要进行部署，第 12 章介绍如何部署微服务分布式项目，以及架构方案、设计方式、编程框架、网关等。

本书适合作为高等学校计算机、软件工程等相关专业 Java 架构课程的教材，也可供对微服务分布式架构感兴趣的人员参考阅读。

本书由张方兴编著。本书即将付梓，还要感谢张鑫编辑持之以恒的支持与帮助。编者微信公众号为北漂程序员的吐槽人生（微信号：beipiaochengxuyuan），读者有任何意见与建议，可随时沟通。

非常感谢阅读本书，学无止境，与君共勉。

由于编者水平有限，加之编写时间仓促，书中难免出现错误与不足之处，欢迎读者批评指正。

编　者
2020 年 1 月

目 录

第1章 微服务分布式架构设计原理 ……1
1.1 Java Web 应用程序的发展历史 ……1
1.2 微服务分布式 ……2
1.2.1 Spring Boot 微服务的定义和特点 ……3
1.2.2 Spring Boot 的职场导读 ……3
1.2.3 Spring 部分内容 ……4
1.2.4 微服务的拆分 ……6
1.3 【实例】微服务工程 Hello World ……7
1.3.1 实例背景 ……7
1.3.2 创建 Maven Project ……7
1.3.3 使用空 Maven Project 模板 ……7
1.3.4 编辑 Maven 坐标定位及工程名 ……8
1.3.5 检查 Maven 目录结构 ……9
1.3.6 编写 Pom 文件 ……10
1.3.7 Spring Boot 依赖包的导入 ……12
1.3.8 编写 Spring Boot 启动类 ……14
1.3.9 编写 Spring Boot 接口 ……14
1.3.10 当前项目结构 ……14
1.3.11 启动工程 ……15
1.3.12 Spring Boot 初始化启动后 ……16
1.3.13 实例易错点 ……16
1.4 Spring Boot 启动类扫描 Bean ……18
1.4.1 @SpringBootApplication 注解 ……18
1.4.2 @ComponentScan 注解 ……20
1.4.3 Spring Boot 扫描其他包下文件 ……20
1.5 【实例】将端口号改成 9090 ……21
1.5.1 实例背景 ……21
1.5.2 创建 application.properties 资源配置文件 ……22
1.5.3 增加资源配置文件中的配置信息 ……23
1.5.4 运行结果 ……23
1.5.5 实例易错点 ……23
1.6 YAML 文件 ……24
1.6.1 YAML 文件简介 ……25
1.6.2 YAML 文件的书写格式 ……25
1.7 【实例】使用 YAML 配置文件 ……25
1.7.1 实例背景 ……25
1.7.2 原 properties 文件 ……25
1.7.3 转换格式后的 YAML 文件 ……26
1.7.4 实例易错点 ……26
1.8 【实例】通过单配置文件让工程适应多应用场景 ……27
1.8.1 实例背景 ……27
1.8.2 更改 application.yml 文件 ……27
1.8.3 更改启动类 ……27
1.8.4 输入启动参数 ……29
1.8.5 运行结果 ……30
1.8.6 实例易错点 ……30
1.9 【实例】通过多配置文件使工程适应多应用场景 ……31
1.9.1 实例背景 ……31
1.9.2 新建 SIT 和 UAT 环境所需资源配置文件 ……31
1.9.3 新建系统资源配置文件 ……31
1.9.4 编写启动类 ……31
1.9.5 当前项目结构 ……32
1.9.6 运行结果 ……32
1.10 微服务配置权重 ……32
1.10.1 资源配置信息类型的权重 ……32
1.10.2 资源配置文件类型的权重 ……33
1.10.3 资源配置文件存在位置与权重解读 ……33
1.11 本章小结 ……34
1.12 习题 ……34

第2章 分布式的注册中心 ……35
2.1 注册中心 ……35

- 2.1.1 Eureka 与 Consul 的区别 ·············· 35
- 2.1.2 Consul 的相关术语 ·············· 37
- 2.1.3 Consul 的安装 ·············· 37
- 2.2 Consul 的常用命令 ·············· 37
  - 2.2.1 consul agent -dev ·············· 38
  - 2.2.2 consul -members ·············· 39
  - 2.2.3 consul leave ·············· 40
  - 2.2.4 agent 命令的常用配置参数 ····· 40
  - 2.2.5 HTTP API ·············· 41
- 2.3 【实例】创建第一个微服务分布式项目 ·············· 42
  - 2.3.1 实例背景 ·············· 42
  - 2.3.2 搭建 Consul 集群 ·············· 42
  - 2.3.3 创建微服务工程编写相应依赖文件 ·············· 45
  - 2.3.4 Spring Cloud 和 Spring Boot 的版本对应关系 ·············· 46
  - 2.3.5 编写微服务 YAML 资源配置文件 ····· 46
  - 2.3.6 编写微服务启动类注册到 Consul 上 ·············· 48
  - 2.3.7 当前项目结构 ·············· 48
  - 2.3.8 运行结果 ·············· 49
  - 2.3.9 实例易错点 ·············· 50
- 2.4 【实例】通过代码获取 Consul 中的服务信息 ·············· 51
  - 2.4.1 实例背景 ·············· 51
  - 2.4.2 编写获得其他注册服务的代码 ····· 52
  - 2.4.3 运行结果 ·············· 53
  - 2.4.4 实例易错点 ·············· 53
- 2.5 【实例】Spring Cloud 操作 Consul 的 K/V 存储 ·············· 54
  - 2.5.1 实例背景 ·············· 54
  - 2.5.2 添加依赖 ·············· 54
  - 2.5.3 利用 Consul 的 UI 界面添加 K/V 存储 ·············· 54
  - 2.5.4 编写 YAML 资源配置文件对应 K/V 存储 ·············· 55
  - 2.5.5 编写 MyConfig.java 文件对应

- 相关 K/V 存储 ·············· 56
- 2.5.6 调用 MyConfig.java 中的参数 ····· 57
- 2.5.7 在启动类引用相关配置 ·············· 57
- 2.5.8 当前项目结构 ·············· 58
- 2.5.9 运行结果 ·············· 58
- 2.5.10 实例易错点 ·············· 60
- 2.6 本章小结 ·············· 60
- 2.7 习题 ·············· 60

第3章 分布式的通信 ·············· 61

- 3.1 分布式通信 ·············· 61
  - 3.1.1 Spring Cloud Feign ·············· 61
  - 3.1.2 Swagger ·············· 61
- 3.2 【实例】微服务集成 Swagger ·············· 62
  - 3.2.1 实例背景 ·············· 62
  - 3.2.2 编写 Swagger 依赖 ·············· 62
  - 3.2.3 编写 Swagger 配置 ·············· 63
  - 3.2.4 编写接口与接口处的 Swagger 配置 ·············· 64
  - 3.2.5 当前项目结构 ·············· 66
  - 3.2.6 运行效果 ·············· 66
  - 3.2.7 实例易错点 ·············· 70
- 3.3 【实例】Feign 调用微服务接口 ·············· 72
  - 3.3.1 实例背景 ·············· 72
  - 3.3.2 引入相关配置信息 ·············· 73
  - 3.3.3 编写 Feign 客户端 ·············· 73
  - 3.3.4 编写调用 ·············· 75
  - 3.3.5 编写启动类 ·············· 76
  - 3.3.6 当前项目结构 ·············· 76
  - 3.3.7 运行结果 ·············· 77
  - 3.3.8 实例易错点 ·············· 77
- 3.4 【实例】Feign 的拦截器 ·············· 78
  - 3.4.1 实例背景 ·············· 78
  - 3.4.2 在 cloud-admin-8084 工程中增加拦截器 ·············· 78
  - 3.4.3 当前项目结构 ·············· 79
  - 3.4.4 运行结果 ·············· 79
  - 3.4.5 实例易错点 ·············· 80

3.5　Feign 的配置 ················· 81
　　3.5.1　传输数据压缩配置 ········ 81
　　3.5.2　日志配置 ··············· 82
　　3.5.3　超时配置 ··············· 83
3.6　【实例】Feign 的降级回退处理
　　——Feign 的 Fallback 类 ······ 84
　　3.6.1　实例背景 ··············· 84
　　3.6.2　在资源配置文件中开启 Feign 内置
　　　　　的 Hystrix 权限 ·········· 84
　　3.6.3　编写 Fallback 降级类 ····· 84
　　3.6.4　Service 整合 Fallback 降级类 ··· 84
　　3.6.5　当前项目结构 ··········· 85
　　3.6.6　运行结果 ··············· 85
3.7　【实例】Feign 的降级回退处理
　　——Feign 的 Fallback 工厂 ···· 86
　　3.7.1　实例背景 ··············· 86
　　3.7.2　编写 Fallback 降级工厂 ··· 86
　　3.7.3　整合 Fallback 降级工厂 ··· 87
　　3.7.4　实例易错点 ············· 87
3.8　本章小结 ····················· 88
3.9　习题 ························· 88

第 4 章　分布式的客户端负载均衡 ··· 89
4.1　负载均衡 ····················· 89
　　4.1.1　传统服务器端负载均衡 ··· 89
　　4.1.2　Ribbon 客户端负载均衡 ··· 89
4.2　【实例】Feign 整合 Ribbon 分发
　　请求 ························· 90
　　4.2.1　实例背景 ··············· 90
　　4.2.2　编写 cloud-book-8086 启动类与
　　　　　配置类支持 Ribbon ······· 91
　　4.2.3　Service 和 Controller ····· 92
　　4.2.4　当前项目结构 ··········· 94
　　4.2.5　运行效果 ··············· 95
　　4.2.6　实例易错点 ············· 96
4.3　Ribbon 的负载均衡策略配置 ··· 97
4.4　本章小结 ····················· 98
4.5　习题 ························· 98

第 5 章　分布式的断路器 ········· 99
5.1　断路器 ······················· 99
　　5.1.1　为什么需要断路器 ······· 99
　　5.1.2　Hystrix ················· 99
　　5.1.3　Hystrix 解决的问题 ····· 100
　　5.1.4　Hystrix 如何解决问题 ··· 100
5.2　【实例】Hystrix 断路器的降级
　　回退 ························ 101
　　5.2.1　实例背景 ·············· 101
　　5.2.2　编写相关 Pom 文件 ····· 101
　　5.2.3　编写 application 资源配置文件 ··· 101
　　5.2.4　编写 Ribbon 配置类 ····· 102
　　5.2.5　编写启动类 ············ 102
　　5.2.6　编写 Service 类 ········· 103
　　5.2.7　编写 Controller 类 ······ 103
　　5.2.8　当前项目结构 ·········· 104
　　5.2.9　运行结果 ·············· 105
　　5.2.10　实例易错点 ··········· 106
5.3　Hystrix 线程池 ··············· 108
　　5.3.1　Hystrix 断路器注解式的命令
　　　　　配置 ·················· 109
　　5.3.2　Hystrix 断路器的注解式线程池
　　　　　配置 ·················· 111
　　5.3.3　Hystrix 断路器注解式的整体
　　　　　定制配置 ·············· 112
　　5.3.4　Hystrix 断路器资源配置式的
　　　　　整体定制配置 ·········· 113
5.4　【实例】Hystrix 断路器的请求
　　缓存 ························ 114
　　5.4.1　实例背景 ·············· 114
　　5.4.2　通过 Filter 初始化 Hystrix
　　　　　上下文 ················ 114
　　5.4.3　让启动类扫描 Filter 过滤器 ··· 116
　　5.4.4　编写 Controller 的 Helper 类 ··· 116
　　5.4.5　编写 Controller 类 ······ 118
　　5.4.6　当前项目结构 ·········· 118
　　5.4.7　运行结果 ·············· 119
　　5.4.8　销毁 Hystrix 的请求缓存 ··· 121

5.4.9 实例易错点 ………………………… 121
5.5 【实例】Hystrix 的请求合并 ……… 123
　　5.5.1 实例背景 …………………………… 123
　　5.5.2 增加@HystrixCollapser 请求合并
　　　　　修饰的函数 …………………………… 124
　　5.5.3 Controller 中调用请求合并函数 … 126
　　5.5.4 当前项目结构 ……………………… 126
　　5.5.5 运行结果 …………………………… 127
　　5.5.6 实例易错点 ………………………… 128
5.6 【实例】Hystrix 的可视化监控 …… 129
　　5.6.1 实例背景 …………………………… 129
　　5.6.2 Hystrix 可视化监控的依赖 ……… 129
　　5.6.3 Hystrix 可视化监控的启动类 …… 129
　　5.6.4 被监控的微服务增加响应地址 …… 130
　　5.6.5 当前项目结构 ……………………… 131
　　5.6.6 运行结果 …………………………… 132
　　5.6.7 实例易错点 ………………………… 134
5.7 本章小结 …………………………… 135
5.8 习题 ………………………………… 135

第 6 章 微服务的异步线程池 …………… 136
6.1 异步线程池 ………………………… 136
　　6.1.1 异步线程池特点 …………………… 136
　　6.1.2 常见的线程池 ……………………… 136
6.2 【实例】创建无返回值异步线
　　　程池 ……………………………… 137
　　6.2.1 实例背景 …………………………… 137
　　6.2.2 编写 Pom 文件 …………………… 137
　　6.2.3 编写 Spring Boot 启动类 ………… 138
　　6.2.4 编写异步线程池任务接口与
　　　　　实现 …………………………………… 138
　　6.2.5 编写外部可调用接口 ……………… 139
　　6.2.6 当前项目结构 ……………………… 140
　　6.2.7 运行程序查看异步线程池效果 …… 140
　　6.2.8 实例易错点 ………………………… 141
6.3 【实例】创建有返回值异步
　　　线程池 …………………………… 141
　　6.3.1 实例背景 …………………………… 141

6.3.2 增加新的服务接口 ………………… 141
6.3.3 增加新的服务实现 ………………… 141
6.3.4 增加新的调用 ……………………… 142
6.3.5 当前项目结构 ……………………… 142
6.3.6 运行程序查看异步线程池效果 …… 142
6.3.7 实例易错点 ………………………… 143
6.4 【实例】优化异步线程池 ………… 143
　　6.4.1 实例背景 …………………………… 143
　　6.4.2 创建初始化线程池配置类 ………… 143
　　6.4.3 更改无返回值的异步线程池
　　　　　Service 实现类 ……………………… 145
　　6.4.4 运行程序查看异步线程池效果 …… 145
　　6.4.5 实例易错点 ………………………… 146
6.5 【实例】优雅停止异步线程池 …… 146
　　6.5.1 实例背景 …………………………… 146
　　6.5.2 何为"优雅" ……………………… 146
　　6.5.3 修改原 Config 配置类 …………… 147
　　6.5.4 修改原 Controller 控制层 ……… 148
　　6.5.5 当前项目结构 ……………………… 149
　　6.5.6 优雅停止异步线程池的执行
　　　　　效果 …………………………………… 150
　　6.5.7 实例易错点 ………………………… 152
6.6 @Enable*注解 …………………… 152
6.7 本章小结 …………………………… 152
6.8 习题 ………………………………… 153

第 7 章 微服务整合持久化数据源 ……… 154
7.1 spring-data ……………………… 154
　　7.1.1 ORM 规范 ………………………… 154
　　7.1.2 JPA、Hibernate、spring-data-jpa
　　　　　之间的关系 …………………………… 155
　　7.1.3 安装 MySQL ……………………… 155
7.2 【实例】Spring Boot 整合 MyBaits
　　　注解式编程 ……………………… 156
　　7.2.1 实例背景 …………………………… 156
　　7.2.2 添加 Pom 文件 …………………… 156
　　7.2.3 编写 application 资源配置文件 … 157
　　7.2.4 编写 dao 层 ……………………… 157

7.2.5	编写访问接口	159
7.2.6	当前项目结构	160
7.2.7	运行效果	160
7.2.8	实例易错点	160

7.3 @Mapper 注解详解 ········ 162

- 7.3.1 @Mapper 和 XML 形式的对应关系 ········ 162
- 7.3.2 MyBatis 的注解式编程多表查询 ········ 162
- 7.3.3 MyBatis 的注解式编程分页查询 ········ 163
- 7.3.4 注册 DataSource 数据源 ········ 165

7.4 【实例】Spring Boot 整合 spring-data-jpa ········ 166

- 7.4.1 实例背景 ········ 166
- 7.4.2 添加 Pom 文件 ········ 166
- 7.4.3 添加资源配置文件中的相关信息 ········ 166
- 7.4.4 添加实体类映射 ········ 167
- 7.4.5 添加 JPA 的 dao 层 ········ 168
- 7.4.6 添加 Controller 控制层查询 JPA 的 dao 层 ········ 169
- 7.4.7 当前项目结构 ········ 170
- 7.4.8 运行结果 ········ 170
- 7.4.9 实例易错点 ········ 170

7.5 本章小结 ········ 171
7.6 习题 ········ 171

第 8 章 微服务事务 ········ 172

8.1 @Transactional 注解 ········ 172

- 8.1.1 @Transactional 声明式事务的传播行为 ········ 173
- 8.1.2 脏读、不可重复读与幻读 ········ 173
- 8.1.3 @Transactional 声明式事务的隔离级别 ········ 174
- 8.1.4 @Transactional 声明式事务的超时时间 ········ 175
- 8.1.5 @Transactional 声明式事务的只读 ········ 175
- 8.1.6 @Transactional 声明式事务指定异常 ········ 176

8.2 【实例】Spring Boot 整合声明式事务 ········ 176

- 8.2.1 实例背景 ········ 176
- 8.2.2 整合@Transactional 的 Service 层编写 ········ 177
- 8.2.3 整合@Transactional 的 Controller 层编写 ········ 178
- 8.2.4 当前项目结构 ········ 179
- 8.2.5 运行结果 ········ 179
- 8.2.6 实例易错点 ········ 180

8.3 本章小结 ········ 181
8.4 习题 ········ 181

第 9 章 微服务的缓存与分布式的消息通信 ········ 182

9.1 Redis ········ 182

- 9.1.1 BSD 协议 ········ 182
- 9.1.2 Java 与 Redis 的历史 ········ 183
- 9.1.3 Spring Data Redis ········ 183

9.2 【实例】微服务整合 Spring Data Redis 增删改查 ········ 184

- 9.2.1 实例背景 ········ 184
- 9.2.2 编写 application.properties 资源配置文件 ········ 184
- 9.2.3 配置 RedisTemplate 模板 ········ 185
- 9.2.4 编写操作 Redis 的工具类 ········ 186
- 9.2.5 编写实体类及接口调用 ········ 188
- 9.2.6 当前项目结构 ········ 189
- 9.2.7 运行结果 ········ 189
- 9.2.8 实例易错点 ········ 190

9.3 【实例】分布式使用 Redis 实现消息通信 ········ 190

- 9.3.1 消息通信应用场景 ········ 190
- 9.3.2 Redis 与 MQ 一系列消息队列的区别 ········ 191
- 9.3.3 实例背景 ········ 191

9.3.4 在 send 微服务中配置模板 ……… 192
9.3.5 在 send 微服务中定时向队列
发布数据 ……………………… 192
9.3.6 在 listener 微服务中编写订阅渠道
的配置信息 …………………… 193
9.3.7 在 listener 微服务中编写监听
实现类 ………………………… 195
9.3.8 当前项目结构 ………………… 195
9.3.9 send 微服务与 listener 微服务运行
结果 …………………………… 196
9.3.10 实例易错点 …………………… 196
9.4 Spring Cache 与 Spring Data Redis
的区别 ……………………………… 196
9.5 【实例】保持 MySQL 与 Redis
数据一致性 ………………………… 197
9.5.1 实例背景 ……………………… 197
9.5.2 编写资源配置文件 …………… 198
9.5.3 编写实体类 Java Bean ……… 198
9.5.4 编写 JPA 仓库 ……………… 199
9.5.5 编写 Service 接口及实现类 … 199
9.5.6 编写 Controller 接口进行测试 … 202
9.5.7 当前项目结构 ………………… 202
9.5.8 运行结果 ……………………… 202
9.5.9 实例易错点 …………………… 203
9.6 本章小结 …………………………… 204
9.7 习题 ………………………………… 204

第 10 章 微服务的任务调度与分布式的
任务调度 …………………………… 205
10.1 【实例】微服务整合任务调度 …… 205
10.1.1 实例背景 …………………… 205
10.1.2 编写任务调度实现类 ……… 205
10.1.3 编写资源配置文件 ………… 206
10.1.4 当前项目结构 ……………… 206
10.1.5 运行效果 …………………… 206
10.1.6 实例易错点 ………………… 206
10.2 @Scheduled 注解详解 …………… 207
10.2.1 cron 表达式 ………………… 207

10.2.2 每个字段允许值 …………… 207
10.2.3 cron 特殊字符意义 ………… 208
10.2.4 常用 cron 表达式 …………… 208
10.3 任务调度的分布式 ……………… 209
10.3.1 任务调度的分布式解决方案 … 209
10.3.2 任务调度的分布式实现原理 … 210
10.4 【实例】微服务整合任务调度
分布式 …………………………… 210
10.4.1 实例背景 …………………… 210
10.4.2 增加 Quartz 依赖 …………… 210
10.4.3 在数据库中增加 Quartz 分布式
的管理表 …………………… 210
10.4.4 编写资源配置文件 ………… 211
10.4.5 创建任务调度管理 Java Bean … 212
10.4.6 创建所需执行的任务 ……… 213
10.4.7 创建执行任务的操作类 …… 213
10.4.8 增加控制层 ………………… 216
10.4.9 当前项目结构 ……………… 218
10.4.10 运行效果 ………………… 218
10.4.11 实例易错点 ……………… 219
10.5 本章小结 ………………………… 220
10.6 习题 ……………………………… 220

第 11 章 微服务的文件上传与分布式
文件管理 …………………………… 221
11.1 文件上传/下载原理 ……………… 221
11.1.1 SpringMVC 文件上传原理 …… 223
11.1.2 文件下载原理 ……………… 225
11.2 【实例】微服务的单文件和
多文件上传 ……………………… 226
11.2.1 实例背景 …………………… 226
11.2.2 编写 application.properties 资
源配置文件 ………………… 227
11.2.3 编写相关接口 ……………… 228
11.2.4 编写前台页面 ……………… 230
11.2.5 当前项目结构 ……………… 231
11.2.6 运行结果 …………………… 232
11.2.7 实例易错点 ………………… 233

11.3 分布式文件管理 235
11.3.1 分布式文件管理特性 235
11.3.2 分布式文件管理解决的问题 235
11.3.3 分布式文件管理解决方案 235
11.4 FastDFS 解决方案 235
11.4.1 FastDFS 的存储策略 236
11.4.2 FastDFS 的文件上传过程 236
11.4.3 FastDFS 的文件同步过程 236
11.4.4 FastDFS 的文件下载过程 237
11.5 FastDFS 的安装部署 237
11.5.1 安装 LibFastCommon 237
11.5.2 安装 FastDFS 237
11.5.3 配置 FastDFS 的跟踪服务器 238
11.5.4 配置 FastDFS 的数据存储服务器 239
11.5.5 配置 FastDFS 的客户端并测试 240
11.5.6 安装 Nginx 部署 FastDFS 240
11.6 【实例】分布式微服务整合 FastDFS 243
11.6.1 实例背景 243
11.6.2 编写 FastDFS 核心配置类 244
11.6.3 编写 FastDFS 工具类 244
11.6.4 编写测试接口 245
11.6.5 当前项目结构 246
11.6.6 运行结果 246
11.6.7 实例易错点 247
11.7 本章小结 248
11.8 习题 248

第 12 章 扩展与部署 249
12.1 微服务分布式架构相关方案总结 249
12.1.1 解决方案与目标 249
12.1.2 分布式部分技术细节扩展 250
12.1.3 动静分离 250
12.1.4 前后端分离 250
12.1.5 数据库读写分离与主从分离 251
12.1.6 应用层与数据层分离 251
12.1.7 CDN 加速 251
12.1.8 异步架构 251
12.1.9 响应式编程 251
12.1.10 冗余化管理 252
12.1.11 灰度发布 252
12.1.12 页面静态化 252
12.1.13 服务端主动推送 253
12.2 微服务扩展 253
12.2.1 微服务整合日志 253
12.2.2 微服务整合单元测试 253
12.2.3 微服务整合全局异常 253
12.2.4 微服务整合 JSR-303 验证机制 254
12.2.5 微服务整合国际化 254
12.2.6 微服务整合安全与认证 254
12.2.7 微服务整合 WebSocket 协议 254
12.2.8 微服务整合 HTTPS 255
12.2.9 微服务整合批处理 255
12.2.10 微服务整合 lombok 255
12.2.11 微服务整合异步消息驱动 255
12.2.12 分布式链路监控 255
12.2.13 分布式单点登录 256
12.3 【实例】分布式网关的初步测试 256
12.3.1 实例背景 256
12.3.2 使用资源配置文件的方式配置分布式网关 256
12.3.3 使用注册 Bean 的方式配置分布式网关 257
12.3.4 运行结果 258
12.4 微服务打包 258
12.4.1 Jar 包 258
12.4.2 War 包 259
12.5 本章小结 259
12.6 习题 259

参考文献 260

第 1 章 微服务分布式架构设计原理

本章按照微服务分布式架构的顺序对其进行讲解，并以实例形式介绍如何编写分布式微服务的代码。每个微服务的架构都先从微服务的注册、微服务间的通信开始，然后编写每个微服务的持久化、缓存等内容。

1.1 Java Web 应用程序的发展历史

Java Web 应用程序的发展历史经历了以下几个过程。

1. Servlet 类项目

在 web.xml 中对 Servlet 容器方式的项目架构进行配置。截至目前依然还有部分老项目用 Servlet 方式进行运行。

2. EJB 类项目

EJB 将业务逻辑从客户端软件中抽取出来，封装在一个组件中。EJB 组件运行在一个独立的服务器上，客户端软件通过网络调用组件提供的服务以实现业务逻辑，而客户端软件的功能是只负责发送调用请求和显示处理结果。在 Java 中，能够实现远程对象调用的技术是 RMI，而 EJB 技术的基础正是 RMI。通过 RMI 技术，J2EE 将 EJB 组件创建为远程对象，客户端可以通过网络调用 EJB 对象。

3. REST+JBoss 平台类项目

REST+JBoss 的设计风格很独特。JBoss 提供了一个编程平台，以及一系列拦截器和 Web 类工具，在编写代码时拦截并过滤模块、向外提供服务，构建 REST 轻量级开发，只需编写很少的配置文件即可引用相关拦截器。在 REST+JBoss 平台时期，国内很多公司也在研发自己的 Java 平台，平台中会集成多种框架和工具，如线程池、持久化、缓存、异步、数据模型等，方便公司内部员工编程。

4. SSM 类项目（Spring+Struts/Struts2+Hibernate/MyBatis）

Java 逐渐开始研发更完善的轻量级开发架构，减少配置文件等压力，将程序专注于业务本身。在 EJB 时期每个项目至少都有几十个配置文件，相比较下 SSM 减轻了依赖很多配置文件的压力，因为 SSM 类项目的架构快速便捷，大部分功能由 Spring 进行集中管理。由于 SSM 架构具备此优点，所以国内部分公司开始减少重复开发的行为，小规模公司则很少再开发公司内部的编程平台。

5．SOA 类项目

在单个 SSM 无法解决大批量高并发需求时，人们开始将一些大型程序分成多个独立运行的模块进行处理，将每个模块部署在不同的服务器上，模块之间用 WebService 协议进行通信。虽然在 SOA 初期便逐渐有了微服务分布式的苗头，但在 SOA 阶段如果大型 Java 项目写得精细一些，每个 WebService 下就都会有相关的管理工具、调度页面、性能监控、业务监控、前置系统和黑白名单，整个大型项目的代码十分复杂且数量庞大。在 SOA 模式下写出来的项目十分安全、稳定、性能优良。因此，时至今日安防类、金融类项目仍然会采取 SOA 架构模式来架构系统，并且在不同的公司之间通信通常也会采用 WebService 协议。

6．Dubbo+Zookeeper 类项目

Dubbo+Zookeeper 分布式服务架构的项目也流行了一段时间，Dubbo 是阿里巴巴公司于 2011 年末开源的一个高性能服务框架，使用高性能 RPC 实现服务的输入和输出，属于分布式项目。在分布式的思想上与 Spring Cloud + Spring Boot 并无任何区别，只是因为 Dubbo 还没有推广起来，阿里巴巴公司就在 2013 年停止了 Dubbo 的更新维护，所以这个服务框架渐渐地被埋没了，虽然现在阿里巴巴公司正在重启 Dubbo 项目，但目前 Spring Cloud 过于火热，Dubbo 一直处于不温不火的状态。

7．Spring Cloud+Spring Boot 类项目

微服务的分布式架构利用 Eureka 或 Consul 平台进行注册，将每个业务分成一个个小的 Spring Boot 微服务包，让每个微服务包独立运行后在注册平台进行注册，通过 Spring Cloud Feign 进行相互通信，也是目前主流的架构。

1.2 微服务分布式

微服务工程将众多框架通过一个入口集成到自身的程序中，让编程人员不需关注框架原理即可直接进行调用。当初的 SSM 类项目也基于相同的思想，但当 SSM 类项目集成的框架过多后，配置文件数目与日俱增且难以管理，而此时 Spring Boot 微服务赋予了它们所需基本配置参数的大部分框架和工具，将初始配置参数的压力再次缩减。

分布式架构将不同的业务模块通过不同的服务器运行，曾经的多个 WebService 组建的 SOA 架构的项目也基于相同的思想，而此时的 Spring Cloud 把过去的架构变得更加简便、轻松。

从 SOA 类项目开始将大型应用程序拆分成多个独立运行的小型应用程序，多个应用程序之间由 WebService 协议进行通信。因为每个 WebService 服务还要集成 SSM，甚至每个服务还要做相关的管理页面和各种监控，所以每个 WebService 服务开发的过程都比较冗长。

微服务架构下的每个微服务和 WebService 一样都是独立运行的，将复杂的业务以模块的形式拆开，方便开发人员编写。微服务通常使用 Spring Boot 进行快捷开发，大部分基础信息在 Spring Boot 的底层之中都已经被配置好了，不像 SSM 一样需要编写大量 XML 程序，因此每个微服务在编程的过程中都会十分快速。

1.2.1　Spring Boot 微服务的定义和特点

Spring Boot 微服务可以轻松地创建独立的产品级应用程序。Spring 平台和第三方库采取自行管理的方式，可更加便捷地开始编写应用程序。Spring Boot 通过 pom.xml 文件，依赖 Spring Boot 的 Starter，将所有需要的 Jar 包集中在一起。Spring Boot 微服务替代了原本的 Spring + SpringMVC/Struts2 + MyBatis/Himbernate 结构。

Spring Boot 做了很多配置文件的管理。原本需要写三四个 Application.xml 文件，对数据库资源池、事务、服务、视图、静态资源等进行配置，现在则完全不需要，而事实上与以前做的传统项目大部分的配置文件也没有较大的区别。

打包变成了直接运行的 Jar 包，因为包变小了，包内所提供的内容更加清晰明确，所以称为微服务，特点如下。

- 创建独立的 Spring 应用程序。
- 直接嵌入 Tomcat、Jetty 或 Undertow（不需要部署 war 文件）。
- 提供被约定好的 Starter 依赖内容，以简化构建配置。
- 尽可能自动配置 Spring 和第三方库。
- 提供产品在生产过程中所需要的如健康检查和外部化配置等。
- 没有绝对的代码生成，也不需要 XML 配置。

Spring Boot 的部署简单，只要服务器有 Java 环境便可运行，即使再多的服务器也能快速且稳定运行。

Spring Boot 的设计初衷是约定大于配置，过多的重复性配置压力在 SSM 里让人深有体会，每个 SSM 项目中的 XML 大部分内容都是类似的，对于资深程序员来讲，任何一个新项目的编写都是一个复制粘贴的过程，而 Spring Boot 省略了初始化配置和重复性配置的过程。

1.2.2　Spring Boot 的职场导读

如今 Spring Boot+Spring Cloud 的微服务分布式项目架构，如同当年的 SSM 类架构一样，属于 1～3 年工作经验 Java Web 研发程序员找工作的必备技能之一。通常初级 Java Web 研发职位的技能要求如下：

（1）熟练掌握 Java 后台编码基础技术，如 JVM、语法、GC、多线程、IO、网络编程、反射等；

（2）熟练掌握 Spring Cloud 与 Spring Boot 架构开发；

（3）熟悉 Spring、SpringMVC、MyBatis 等开源框架，最好了解其中部分源码；

（4）熟悉 Tomcat、JBoss、Weblogic 等主流应用服务器；

（5）熟悉 HTML、CSS、JavaScript、JQuery 等前段基础知识；

（6）熟悉 Oracle 或 MySQL 数据库开发，如 SQL 与存储过程；

（7）熟练使用 Linux 操作系统，掌握基本 Shell 命令；

（8）熟练掌握 Radis、MongoDB；

（9）熟练使用 WebService 主流框架，如 CXF、Axis 等；

（10）熟练使用 Maven、Git、SVN 等相关版本控制工具。

目前，主流的架构通常用 Spring Boot + Spring Cloud 作为程序后台，用 VUE 或 AngrulJS 作为程序前台，达到 VUE+Spring Boot+Spring Cloud 前后台分离的架构模式，Java 程序员再不需要关心 JSP 等前台相关内容，这些都由前端人员自行管理，减轻了沟通成本，也减轻不少编程和维护上的压力。

微服务类框架除了 Spring Boot，还有 JFinal、JBoot 等国产微服务框架。

1.2.3 Spring 部分内容

如表 1-1 所示，Spring 的常用框架有很多种，而且 Spring 定制了大多分布式架构所需的工具与框架。Spring 已经替编程人员完成了分布式的架构，而分布式编程人员在其中挑选一些适合项目的工具并参照操作文档进行集成即可。

表 1-1 Spring 的常用框架

主 项 目	子 项 目	作 用
Spring Framework		Spring 框架核心 IOC/AOP
Spring Boot		微服务
Spring Data	Spring Data JDBC	通过模板的方式操作持久化数据库，但是不提供缓存、延迟加载等
	Spring Data JDBC Extensions	Spring Data JDBC 的扩展框架，提供了 Oracle 数据库的一些高级功能，包括故障转移、高级队列、高级数据类型、数据源预置器等
	Spring Data JPA	通过定义函数名或注解的形式使用 JPA 仓库的一种操作持久化数据库的 ORM 类方案
	Spring Data LDAP	使用 Spring 应用程序操控轻量级目录访问协议（LDAP）的实现方案
	Spring Data MongoDB	通过模板方式操作非关系型数据库 MongoDB 的方案，编写方式类似于 Spring Data JDBC
	Spring Data Redis	通过模板方式操作非关系型数据库 Redis 的方案，编写方式类似于 Spring Data JDBC
	Spring Data R2DBC	JDBC 驱动程序的反应变体称为 R2DBC，它允许数据异步流式传输到已订阅它的任何端点，结合使用 R2DBC 等反应式驱动程序；类似于异步访问持久化的一种方式，与 Spring WebFlux 能够编写一个完整的响应式应用程序来异步进行数据的接收和发送
	Spring Data REST	使在 Spring 数据存储库上构建超媒体驱动的 REST Web 服务变得更加容易
	Spring Data for Apache Cassandra	最初由 Facebook 公司开发的 Cassandra 是一个开源分布式 NoSQL 数据库系统
	Spring Data for Apache Geode	ApacheGeode 项目的主要目标是使用 ApacheGeode 为分布式数据管理构建高可伸缩性的 Spring 支持的应用程序
	Spring Data for Apache Solr、Spring Data for PivoGemfire、Spring Data Couchbase、Spring Data Elasticsearch、Spring Data Evers、Spring Data Neo4J、Spring for Apache Hadoop 为 Spring Data 子项目里操作不同的数据源，有兴趣可自行查阅相关资料	

续表

主项目	子项目	作用
Spring Cloud	Spring Cloud Config	在所有环境中为应用程序管理外部属性，每个微服务不需要重新定义其环境变量，Spring Cloud Config 允许统一在 GIT、SVN 中进行管理
	Spring Cloud Stream	用于构建与共享消息系统相连接的高度可伸缩的事件驱动微服务
	Spring Cloud for Amazon Web Services	简化了与托管的 Amazon Web 服务的集成，提供了一种使用众所周知的 Spring 习惯用法和 API（如消息传递或缓存 API）与 AWS 提供的服务交互方法，开发人员围绕托管服务构建应用程序，而不必关心基础设施或维护； Amazon Web Services 即 AWS，是 Amazon 公司的云计算 IaaS 和 PaaS 平台服务
	Spring Cloud Bus	将分布式系统的节点与轻量级消息代理连接起来，然后用于广播状态更改（如配置更改）或其他管理指令，目前唯一的实现是使用 AMQP 代理作为传输，但其他传输路线图上有相同的基本功能集； AMQP 即 Advanced Message Queuing Protocol，一个提供统一消息服务的应用层标准高级消息队列协议，是应用层协议的一个开放标准，为面向消息的中间件设计
	Spring Cloud Consul	类似于 Spring Eureka 的注册中心
	Spring Cloud Eureka	分布式的注册中心，每个微服务会在 Eureka 中进行注册，方便微服务之间互相调用，但是 Eureka 2.0 以后已经停止更新
	Spring Cloud Gateway	一种简单而有效的方法使路由到 API，通常给外部请求（非分布式微服务下项目）进行路由
	Spring Cloud Feign	分布式微服务之间互相调用的工具
	Spring Cloud Ribbon	分布式微服务的客户端负载均衡，通过内部进行分发请求，提供了一些负载均衡的策略算法
	Spring Cloud Zookeeper	分布式微服务绑定 Zookeeper 的工具
	Spring Cloud Kubernetes	分布式微服务整合 K8S
	Spring Cloud Task	提供云端管理任务调度
	Spring Cloud Security	分布式权限
	Spring Cloud Zuul	为微服务集群提供代理、过滤、集群的工具
	此处省略了众多不常用的子项目，有兴趣可自行查阅相关资料	
Spring Cloud Data Flow		实时数据处理的工具
Spring Security	Spring Security SAML	身份验证与授权
	Spring Security Oauth	单点登录 SSO
	Spring Security Kerberos	对接 Kerberos 协议
	Spring Security 是 Spring 负责安全部分的框架，其下也有 Spring Security Messaging 为 WebSocket 提供安全、Spring Security Webflux 为异步 Web 提供安全等相关模块，有兴趣可自行查阅相关资料	

续表

主 项 目	子 项 目	作 用
Spring Session	Spring Session MongoDB	分布式登录后的缓存 Token 类工具,提供了一个 API 及实现,用于通过 Spring Data MongoDB 来管理存储在 MongoDB 中的用户会话信息
Spring Web Flow		异步 Web 框架
Spring Web Service		对外提供 Web Service 协议服务
Spring 下还包括以下工具等,因篇幅有限,不再赘述:		
Spring Integration	Spring HATEOAS	Spring RESTDocs
Spring Batch	Spring IO Platform	Spring AMQP
Spring for Android	Spring CredHub	Spring for Apache Kafka
Spring LDAP	Spring Mobile	Spring Roo
Spring Shell	Spring Statemachine	Spring Test HtmlUnit

1.2.4 微服务的拆分

微服务将整个大型 Web 项目拆分成了多个小型的 Spring Boot 工程,彼此依靠 Spring Cloud 相关分布式框架工具进行整合,通常编程人员将一个大项目中不同的业务和技术分别拆分成不同的小模块进行独立运行。例如,一个已上线的项目一般包含以下微服务工程。

(1) zfx_account 微服务工程:关于账号、鉴权、缓存的微服务。

(2) zfx_admin 微服务工程:微服务自身的控制管理服务。

(3) zfx_service 微服务工程:给外部系统提供服务的微服务。

(4) zfx_control 微服务工程:业务监控、性能监控的微服务。

(5) zfx_mobile 微服务工程:给手机端提供接口的微服务。

(6) zfx_produce 微服务工程:管理自身产品的微服务。

(7) zfx_message 微服务工程:邮件、短信通知的微服务。

(8) zfx_pay 微服务工程:对接支付接口的微服务。

(9) zfx_mq 微服务工程:给其他微服务提供消息队列功能的微服务。

(10) zfx_common 微服务工程:通常实体类都存在 common 中,别的项目会依赖 common, common 中存放其他微服务共通性的工具。

(11) zfx_parent 微服务工程:别的项目会依赖 parent,通常在 parent 中不会编写实际意义的代码,parent 主要定义相关框架的版本信息等相关内容。

(12) zfx_paytran 微服务工程:支付相关的微服务。

(13) zfx_dataaccess 微服务工程:一些数据库增删改查外的索引管理、数据库心跳及预警等相关功能的微服务。

上述拆分中每个微服务项目都是一个独立运行的 Spring Boot 工程。每个 Spring Boot 工程之间依靠注册中心进行关联,依靠 Feign 进行相互通信,依靠 Spring Cloud Security 进行鉴权和单点登录。

不同的业务场景所需要进行拆分的业务模块和技术模块肯定不同，通常根据自身的业务进行调整，不过像 zfx_parent、zfx_common 肯定是系统必备的内容。因为一个工程中再多的业务也必须有互相需要依赖的 Jar 包、提取的共同点、共同需要的技术内容，所以在微服务拆分时可先将 parent 和 common 两个模块拆分出来，抽出程序的共通性，再慢慢新增其他的服务模块。

线上项目拆分成 40～50 个微服务工程是正常现象，因此在做分布式微服务项目架构时，只要维护管理到位，就不用担心微服务的数目是否过多，如果重用性得当，让每个程序员管理其中几个微服务，就能达到以最快速度开发迭代的目的。

移动端使用 H5，再利用 Android 和 iOS 的 Webview 整合 H5 进行混合开发，大部分内容在 Android 与 iOS 之间都不需要重新编写，即一套页面在两个系统中使用，不需要重复开发。

1.3 【实例】微服务工程 Hello World

1.3.1 实例背景

本书采用 Eclipse 配合 Maven 进行编程，JDK 采用 1.8 版本，如果使用 Idea、STS 等相应开发工具，并无任何区别。STS 是专门为 Spring Boot 制作的开发工具，全称为 Spring Tool Suite，其中含有各种 Spring Boot 模板，在引入模板时，会自动生成 pom.xml 中所需要依赖的文件，但是本书使用 Eclipse 进行开发。STS 基于 Eclipse 并进行了一定程度的封装，因此其快捷键、代码提示等功能和 Eclipse 无任何区别。

建议 Spring Boot 的版本在 2.0.0 上，因为有一些新特性可以使用。

本实例将创建 boot_01 工程，首先使用 Maven 插件集成 Spring Boot 微服务依赖，再通过@RestController 注解编写 REST 风格的 HTTP 接口输出 Hello World 字符串。

1.3.2 创建 Maven Project

在 Eclipse 工具下进行编程，Spring Boot 通常在 Maven 工具的辅助下创建，因为 Spring Boot 内部所继承的依赖 Jar 包太多，Maven 能够一次性将 Spring Boot 所有需要的依赖 Jar 包下载完成。

因为微服务工程只使用 Spring Boot 而并非 Spring Boot 下的一个模块，所以创建的是 Maven Project，而不是 Maven Module。表 1-2 所示为 Maven 工程类型及其含义。

表 1-2 Maven 工程类型及其含义

Maven 工程类型	含 义
Maven Project	创建独立运行的 Maven 工程
Maven Module	创建某 Maven 工程下的一个模块，其实际存在地址还在另一个 Maven 工程中

1.3.3 使用空 Maven Project 模板

模板的最大优点是能自动生成 pom.xml 文件。例如，当需要引入 MySQL 的驱动包对数据库增删改查时，STS 创建工程可进行选择和引入。

如图 1-1 所示，创建 Maven Project 工程后选中 Create a simple project(skip archetype selection)复选框，即不使用 Maven 模板。

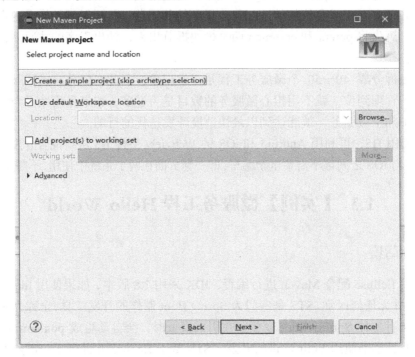

图 1-1

1.3.4 编辑 Maven 坐标定位及工程名

Maven 的一个核心作用是对管理工程的依赖，同时引入所需要的各种 Jar 包等。为了能自动化地解析任何一个 Java 构件，Maven 必须将依赖 Jar 包或其他资源进行唯一标识，标识即 Maven 坐标。Maven 坐标是对工程依赖的基础。拥有其他工程的 Maven 坐标才能引用其他工程，或被其他工程所引用。Maven 坐标的作用好像 Maven 世界中的身份证一样。

Maven 坐标通过 Group Id、Artifact Id、Version、Packaging、Name、Decription 等元素来定义，如同坐标轴中的 X 轴、Y 轴一样。

Group Id：定义当前 Maven 工程隶属的实际项目。图 1-2 中建立的 Jar 文件默认将处于 C:\Users\Administrator\.m2\repository\org\zfx\boot 文件夹中，在 Maven 世界里别人依赖此工程，也会处于$Maven_Home\.m2\repository\org\zfx\boot 文件夹中。通常命名会表明该工程处于各项目中的哪个位置，以方便理解。

Artifact Id：定义 Group Id 位置下实际工程的名字，推荐与工程名保持一致，Maven 生成的构件其文件名都会以 Artifact Id 作为开头。

Version：定义当前 Maven 工程所在的版本，因为作为名字后缀存储，所以建议尽量简短，方便后续对其进行维护。

Packaging：定义 Maven 的打包方式，通常为 Jar 文件或 War 文件。若不对其进行定义，则默认为 Jar 文件。

Decription：对 Maven 工程进行描述的内容，通常不会使用。

图 1-2 所示为输入该工程的 Maven 坐标等基本信息。

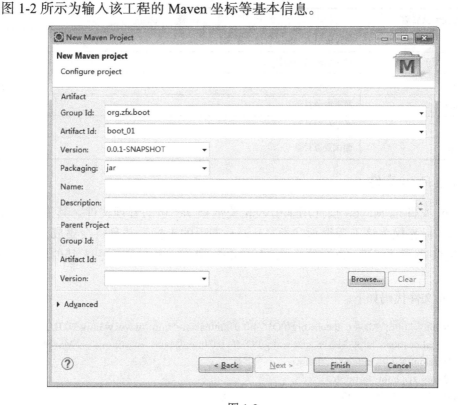

图 1-2

1.3.5 检查 Maven 目录结构

好的目录结构方便开发人员编写程序，也将为程序在未来使用过程中的维护工作打下良好的基础。

Maven 目录结构是业界公认的最佳目录结构，为开发者提供了缺省的标准目录模板，Eclipse 中 Maven 目录结构如图 1-3 所示。Idea 的 Maven 目录结构与 Eclipse 的 Maven 目录结构略有不同。

Maven 目录结构释义如表 1-3 所示。

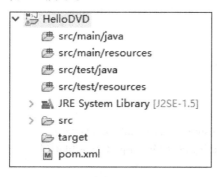

图 1-3

表 1-3　Maven 目录结构释义

文件/文件夹名称	释　　义
pom.xml	Maven 引入其他工程的依赖文件
target	打包后输出的根目录
/src/main/java	源代码目录
/src/main/resource	所需资源目录
/src/test/java	测试代码目录
/src/test/resource	测试资源目录

1.3.6　编写 Pom 文件

pom.xml 文件是 Maven 插件用来引入相关依赖 Jar 包的管理文件。其中，pom.xml 内的<parent>标签代表该工程继承父类 Maven 工程的坐标，并且将会集成父类 Maven 下所有相关依赖，类似于 Java 的 extend。<dependencies>标签下引入其他相关 Maven 工程坐标。

pom.xml 文件代码如下。

```xml
<projectxmlns="http://Maven.apache.org/POM/4.0.0"xmlns:xsi="http://www.w3.org/2001/XMLSchema-instance"xsi:schemaLocation="http://Maven.apache.org/POM/4.0.0 http://Maven.apache.org/xsd/Maven-4.0.0.xsd">
<modelVersion>4.0.0</modelVersion>
<groupId>org.zfx.boot</groupId>
<artifactId>boot_01</artifactId>
<version>0.0.1-SNAPSHOT</version>
    <!-- 继承默认值为 Spring Boot  -->
    <parent>
        <groupId> org.springframework.boot </groupId>
        <artifactId> spring-boot-starter-parent </artifactId>
        <version> 2.0.4.RELEASE </version>
    </parent>
    <!-- 添加 Web 应用程序的典型依赖项 -->
    <dependencies>
        <dependency>
            <groupId> org.springframework.boot </groupId>
            <artifactId> spring-boot-starter-web </artifactId>
        </dependency>
    </dependencies>
    <!-- 作为可执行的 Jar 包  -->
    <build>
        <plugins>
            <plugin>
                <groupId> org.springframework.boot </groupId>
                <artifactId> spring-boot-Maven-plugin </artifactId>
            </plugin>
        </plugins>
```

```
            </build>
</project>
```

更新 Pom 文件前要了解当前计算机上是否含有 JDK1.8 版本，JDK1.7 或以下版本需尝试使用 spring-boot-starter-parent，版本为 1.5.3.RELEASE。而 JDK1.7 使用 Spring Boot 2.0 以上版本会报 Pom 文件相关编译错误。具体可查询 Spring.io 网站信息。

编写本书时 Spring Boot 最高版本为 2.0.x。

1．若无法从 parent 里进行继承

若工程有其他的包需要继承，则直接在 pom.xml 文件中对 Spring Boot 进行引入，即不使用<parent></parent>标签，pom.xml 文件部分代码如下。

```
            <properties>
                    <spring-boot.version>2.0.0.RELEASE</spring-boot.version>
            </properties>
            <dependencyManagement>
                    <dependencies>
                            <dependency>
                                    <groupId>org.springframework.boot</groupId>
                                    <artifactId>spring-boot-dependencies</artifactId>
                                    <version>${spring-boot.version}</version>
                                    <type>pom</type>
                                    <scope>import</scope>
                            </dependency>
                    </dependencies>
            </dependencyManagement>
            <dependencies>
                    <dependency>
                            <groupId>org.springframework.boot</groupId>
                            <artifactId>spring-boot-starter-web</artifactId>
                    </dependency>
            </dependencies>
            <build>
                <plugins>
                    <plugin>
                        <groupId>org.apache.Maven.plugins</groupId>
                        <artifactId>Maven-compiler-plugin</artifactId>
                        <version>3.1</version>
                        <configuration>
                            <source>1.8</source>
                            <target>1.8</target>
                        </configuration>
                    </plugin>
                </plugins>
            </build>
```

注意，如果没有编写<dependencyManagement></dependencyManagement>标签进行定义，经常会报错。Spring Boot 官方文档同样编写了该标签。

2. 保存 pom.xml 文件会报错误

在 Maven 工程的 pom.xml 文件被更改并保存的情况下，工程会报如下错误：

> Project configuration is not up-to-date with pom.xml. SelectMaven->Update Project... from context menu or use Quick Fix...

工程报错效果如图 1-4 所示。该错误是未更新 Maven 工程依赖的报错，原因是虽然进行了依赖但并未导入 Maven 相关依赖包。此时需要对 Maven 工程进行 Update 操作。

1.3.7　Spring Boot 依赖包的导入

Maven Update Project 是指 Maven 进行了一系列跟自身配置有关的操作，如更新 SVN、更新工程关联、编译工程、构建发布、下载 Jar 包、下载 Jar doc 文件等。

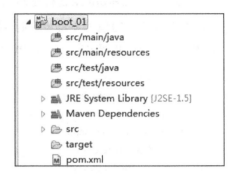

图 1-4

如图 1-5 所示，选择 Update Project 更新 Maven 工程依赖，可解决图 1.4 中在编写完 pom.xml 相关依赖并保存时未更新 Maven 工程的错误。

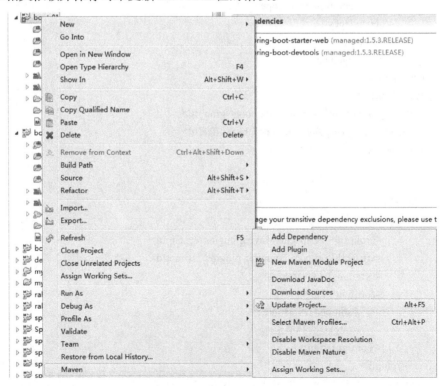

图 1-5

图 1-6 中 Maven Update 导入文件依赖包的选项释义如下。

Offline：启动离线模式，即不会通过 Maven 镜像地址下载资源，而重新检查自身仓库内的资源，内网开发的公司若没有自身的 Maven 镜像，则经常会使用离线模式。

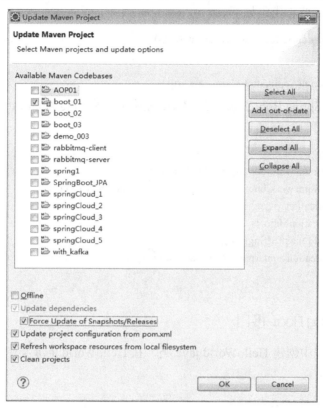

图 1-6

Force Update of Snapshots/Releases：强制更新模式，在 Update dependencies 更新依赖下，强制更新模式代表无论此时所有依赖包是否处于正常依赖状态，都会重新到 Maven 仓库中进行检查，并重新把该包导入工程中。若工程在导入包后仍然发生错误，有些包没有被正常导入，则需要启动强制更新模式进行强制重新导入。

如果编写了一个 zfx_common-1.0.jar 包，已经导入工程中，但是编程人员手动在本地仓库中删除了该依赖包，并置入一个更新后的 zfx_common-1.0.jar 依赖包（但两个版本拥有相同的名称和版本号），就需要启动离线模式加强制更新模式，离线模式加强制更新模式会在自身的工程中删除原有 zfx_common-1.0.jar 包，并依赖更新后的同版本号的 zfx_common-1.0.jar 依赖包。

Update project configuration from pom.xml：根据 pom.xml 的配置文件更新工程。

Refresh workspace resources from local filesystem：从本地仓库系统刷新工作区资源。若仓库中含有该依赖包，则从仓库中进行获取；若仓库中没有该依赖包，则会通过 Maven 镜像地址进行下载。镜像地址在$Maven_HOME/config/setting.xml 文件中配置。

Clean projects：重新编译工程。

1.3.8 编写 Spring Boot 启动类

Spring Boot 微服务工程都是 JavaSE 工程，因此需要 main 方法来启动并加载 Spring 容器等相关操作。启动核心代码如下。

```
New SpringApplicationBuilder().web(true).run(args);
```

和

```
SpringApplication.run(ApplicationMain.class,args);
```

上述两种都是 Spring Boot 微服务的启动核心代码。启动类 Application- Main.java 代码如下（启动类名称可自行定义）。

```java
package com.zfx;
import org.springframework.boot.SpringApplication;
import org.springframework.boot.autoconfigure.SpringBootApplication;
@SpringBootApplication
Public class ApplicationMain {
    public static void main(String[] args) {
        SpringApplication.run(ApplicationMain.class,args);
    }
}
```

1.3.9 编写 Spring Boot 接口

在 boot_01 工程中创建 HelloWorld.java 类，在 HelloWorld.java 类中编写 REST 风格的 HTTP 接口代码如下。

```java
package com.zfx.controller;
import org.springframework.web.bind.annotation.RequestMapping;
import org.springframework.web.bind.annotation.RestController;
@RestController
public class HelloWorld {
    @RequestMapping("/")
    public String hello() {
        return "HelloWorld";
    }
}
```

@RestController 注解相当于@Controller 注解和@ResponseBody 注解的集成注解，@RestController 注解可直接返回 json 格式字符串，不需要新增其他代码。

@RequestMapping("/")是 springMVC 的注解，映射路径给外部进行访问。若该工程中含有相同的@RequestMapping 路径，则会发生启动报错异常。

1.3.10 当前项目结构

ApplicationMain 放置在 com.zfx 文件夹中，而其他文件则放置在 com.zfx.xxx 中，原因是 Spring Boot 的@SpringBootApplication 只能扫描 com.zfx 的下级文件夹。

若某个@Component 组件类注解放置在 com.zfx 同级文件夹下，则无法被@SpringBootApplication 注解扫描到，也无法被注册到 Spring 的 Application 容器内。当前项目结构如图 1-7 所示。

1.3.11 启动工程

因为 Spring Boot 将 Tomcat 集成到自身上的框架，所以通过 main 命令可启动。

启动项目时直接在 ApplicationMain.java 类中运行 Run As→Java Application 命令即可，如图 1-8 所示。

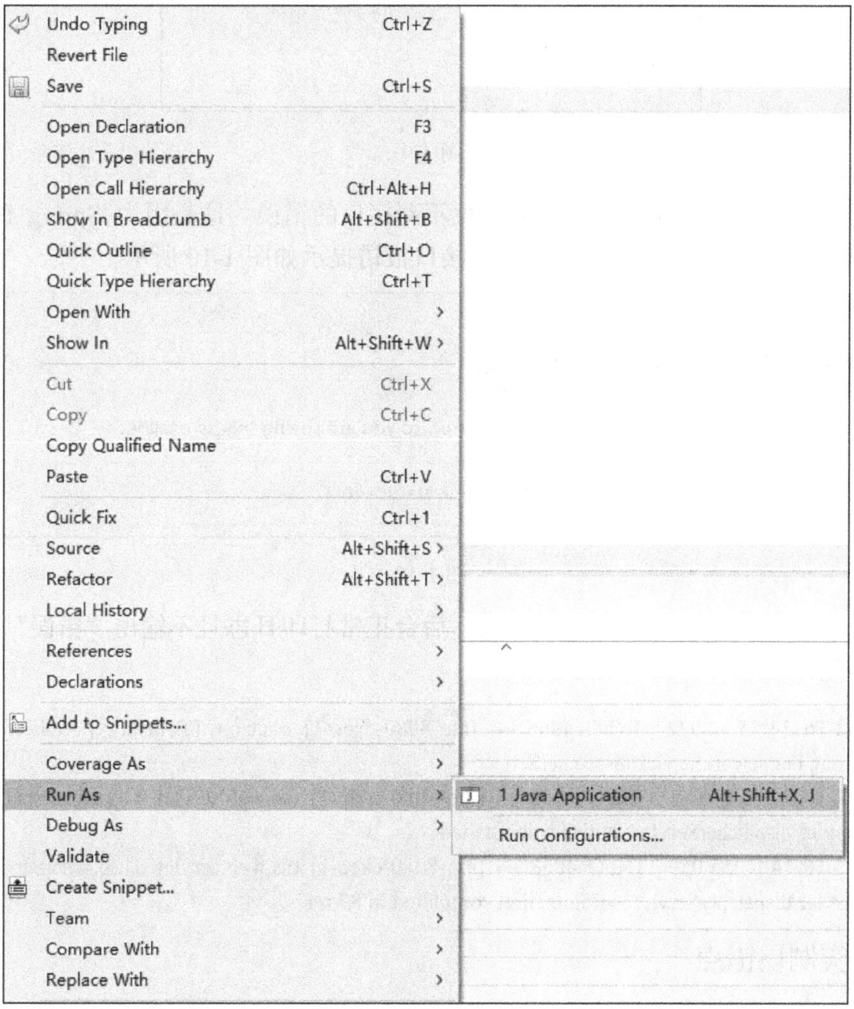

图 1-7

图 1-8

1.3.12　Spring Boot 初始化启动后

根据日志最后看到部分启动信息如下。

```
Tomcat started on port(s): 8080 (http) with context path ''
```

Spring Boot 内嵌 Tomcat 默认加载端口号为 8080。

Spring Boot 默认不需要输入工程名。如果有特殊需要，可在配置文件中增加自定义工程名配置。如果@RequestMapping 写成@RequestMapping("HelloWorld")，访问的路径为 localhost:8080/HelloWorld，后续对工程使用 Spring Cloud 进行控制时可以用增加工程名的方式来区分不同的微服务。

启动工程后，输入 localhost:8080 调用工程中的相关接口，执行结果如图 1-9 所示。

图 1-9

如果访问 localhost:8080/XXX，会报找不到接口的错误。这是因为 Spring Boot 不像 Tomcat 自带了一个欢迎页面。找不到相关接口报错提示如图 1-10 所示。

图 1-10

注意，HTTP-404 错误只在前台报错，后台正常打印日志且不输出"错误"信息，错误日志如下。

```
    2018-08-16 14:05:54.972  INFO 4868 --- [nio-8080-exec-1] o.a.c.c.C.[Tomcat].[localhost].[/]       : Initializing Spring FrameworkServlet 'dispatcherServlet'
    2018-08-16 14:05:54.972  INFO 4868 --- [nio-8080-exec-1] o.s.web.servlet.DispatcherServlet        : FrameworkServlet 'dispatcherServlet': initialization started
    2018-08-16 14:05:55.059  INFO 4868 --- [nio-8080-exec-1] o.s.web.servlet.DispatcherServlet        : FrameworkServlet 'dispatcherServlet': initialization completed in 87 ms
```

1.3.13　实例易错点

1．缺少启动注解错误

SpringContext 容器没有启动成功的报错信息如下，通常是因为缺少@EnableAuto-

Configuration 注解所引起的,此时应检查 main 方法的类是否含有@EnableAutoConfiguration 注解或@SpringBootApplication 注解修饰。

> org.springframework.context.ApplicationContextException: Unable to start embedded container; nested exceptionis org.springframework.context. ApplicationContextException: Unable to start EmbeddedWebApplication Context due to missing EmbeddedServletContainerFactory bean.

2．@RequestMapping 路径重复错误

@RequestMapping 路径重复错误是初学者常见错误之一，错误日志如下。

> org.springframework.beans.factory.BeanCreationException: Error creating bean with name 'requestMapping HandlerMapping' defined in class path resource [org/springframework/boot/autoconfigure/web/WebMvcAuto-Configuration$EnableWebMvcConfiguration.class]: Invocation of init method failed; nested exception is java.lang.IllegalStateException: Ambiguous mapping. Cannot map 'com.zfx.ApplicationMain$Controller' method

通常说明@RequestMapping("")在映射路径时，如果某一个路径相同，重复了一次或以上，需要修改成不同的路径名。

注意，下述代码使用@PathVariable 注解用来获取 URL 路径上携带参数的信息，虽然@RequestMapping(/HelloWorld/{id})注解中的花括号内使用了不同的命名方式，但由于 URL 路径重复，所以@RequestMapping 注解无法正常识别并获取 URL 路径与其携带的参数，无法正确将 URL 映射到 Java 接口中去，因此会引发多个相同路径的错误异常。

```
@RequestMapping("/HelloWorld/{id}")
public String hello1(@PathVariable String id){
    return"";
}
@RequestMapping("/HelloWorld/{name}")
public String hello2(@PathVariable String name){
    return"";
}
@RequestMapping("/HelloWorld/{password}")
public String hello3(@PathVariable String password){
    return"";
}
```

以上方法并不会报启动错误。部分启动日志如下。

> 2018-08-16 14:13:14.771 INFO 6352 --- [main] s.w.s.m.m.a.RequestMappingHandlerMapping: Mapped "{[/HelloWorld/{password}]}" onto public java.lang.String com.zfx.ApplicationMain$Controller.hello3 (java.lang.String)
> 2018-08-16 14:13:14.773 INFO 6352 --- [main] s.w.s.m.m.a.RequestMappingHandlerMapping : Mapped "{[/HelloWorld/{name}]}" onto public java.lang.String com.zfx.ApplicationMain$Controller.hello2 (java.lang.String)
> 2018-08-16 14:13:14.773 INFO 6352 --- [main] s.w.s.m.m.a.RequestMappingHandlerMapping : Mapped "{[/HelloWorld/{id}]}" onto public java.lang.String com.zfx.ApplicationMain$Controller.hello1 (java.lang.String)

多个相同路径的错误在运行后调用会报以下错误，找不到相对应的接口函数如图 1-11 所示。因此应当避免出现多个相同位置的@PathVariable，让维护人员混淆或模糊的写法。

2018-08-16 14:19:59.819 ERROR 6572 --- [nio-8080-exec-2] o.a.c.c.C.[.[. [/].[dispatcherServlet]   : Servlet.service() for servlet [dispatcherServlet] in context with path [] threw exception [Request processing failed; nested exception is java.lang.IllegalStateException: Ambiguous handler methods mapped for HTTP path 'http://localhost:8080/HelloWorld/123452': {public java.lang. String com.zfx.ApplicationMain$Controller.hello1(java.lang.String), public java.lang.String com.zfx.ApplicationMain$Controller.hello3(java.lang.String)}] with root cause

java.lang.IllegalStateException: Ambiguous handler methods mapped for HTTP path 'http://localhost:8080/HelloWorld/123452': {public java.lang.String com.zfx.ApplicationMain$Controller.hello1(java.lang.String), public java.lang. String com.zfx.ApplicationMain$Controller.hello3(java.lang.String)}

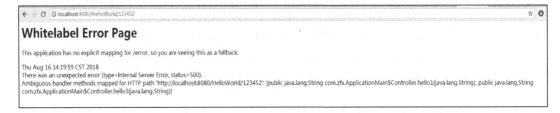

图 1-11

1.4 Spring Boot 启动类扫描 Bean

Spring Boot 默认扫描当前包下级别的包与文件，本节将详细介绍其原因。

1.4.1 @SpringBootApplication 注解

@SpringBootApplication 注解的目的是整合一个或多个配置类及配置信息，其内部集成的注解如下。

```
@Target(ElementType.TYPE)
@Retention(RetentionPolicy.RUNTIME)
@Documented
@Inherited
@SpringBootConfiguration
@EnableAutoConfiguration
@ComponentScan(excludeFilters = {
        @Filter(type = FilterType.CUSTOM, classes = TypeExcludeFilter.class),
        @Filter(type = FilterType.CUSTOM, classes = AutoConfigurationExclude- Filter.class)})
public @interface SpringBootApplication {
```

@ComponentScan 注解帮助 Spring boot 扫描当前工程，让其得到相关的@Controller、@Service、@Configuration、@Component 注解，并对其进行注册为 Java Bean 以供系统调用。

@Target 注解用于设定注解的适用范围，可设置的取值及范围如表 1-4 所示。

表 1-4 @Target 注解的取值及范围

取值	范围
ElementType.METHOD	用于方法上
ElementType.TYPE	用于类或接口上
ElementType.ANNOTATION_TYPE	用于注解类型上（被@Interface 修饰的类型）
ElementType.CONSTRUCTOR	用于构造方法上
ElementType.FIELD	用于域上
ElementType.LOCAL_VARIABLE	用于局部变量上
ElementType.PACKAGE	用于记录 Java 文件的 package 信息
ElementType.PARAMETER	用于参数上

@Retention 注解定义了该 Annotation 被保留的时间,保留时间长短不影响 class 的执行，因为 Annotation 与 class 在使用上是被分离的，@Retention 注释的取值及释义如表 1-5 所示。

表 1-5 @Retention 注解的取值及释义

取值	释义
RetentionPolicy.SOURCE	源文件保留
RetentionPolicy.CLASS	class 保留
RetentionPolicy.RUNTIME	运行时保留

@Documented 注解的作用是给类添加注释，在生成 doc 文档时发挥作用。

@Inherited 注解表示该注解会被子类继承，但成员属性、方法并不受此注释的影响。对于类来说，子类要继承父类的注解需要该注解被 @Inherited 标识。对于成员属性和方法来说，非重写的都会保持和父类一样的注解，而被实现的抽象方法、被重写的方法都不会有父类的注解。

@SpringBootConfiguration 注解的作用是集成相关配置文件。

@EnableAutoConfiguration 注解会根据已添加的 Jar 依赖，对 Spring 容器进行默认配置。

@ComponentScan 注解给 Spring Boot 提供扫描功能，详解参见 1.4.3 小节。

@Filter 注解是 Spring 给 Servlet 容器中的过滤器（Filter）提供注册与实现能力的 Bean。Spring Boot 集成 Servlet 时也经常使用@ComponentScan 注解,@Filter 将以书写 Spring Bean 的方式去编写原本存在 web.xml 的 Filter 信息。@Filter 注解中 type 取值及释义如表 1-6 所示。

表 1-6 @Filter 注解中 type 的取值及释义

取　值	释　义
FilterType.ANNOTATION	注解类型
FilterType.ASSIGNABLE_TYPE	注解指定的类型
FilterType.ASPECTJ	Aspectj 的表达式
FilterType.REGEX	正则表达式
FilterType.CUSTOM	自定义规则

1.4.2　@ComponentScan 注解

Spring Boot 默认扫描当前包下级别的包与文件，如图 1-12 所示的工程，因为 ApplicationMain.java（启动类的名字是自定义的）启动类在 com.zfx 包下，而@Controller 在 com.zfx.controller 包下，所以 Spring Boot 根据默认配置扫描文件并将其内的函数注册为 Spring Bean，包括接下来所需的 com.zfx.service、com.zfx.service.impl、com.zfx.bean、com.zfx.task 等包都不会有任何扫描和注册上的问题。

如果代码在 com.zhang.controller 包下，因为 com.zhang.controller 包不在 com.zfx 包下，所以会报 java.lang.NoSuchMethodException 等找不到 Bean 或类名的错误。如果希望工程找到 com.zhang.controller 包并对其处理，就需要在启动类使用@ComponentScan 注解引入 com.zhang/com.zhang.controller 包。com.zhang 是 Spring Boot 默认的扫描机制与扫描地点。如果启动类在 com.zhang 包内，就扫描到 com.zhang.controller。

为了保证 Spring Boot 启动类扫描整个工程所有的包，启动类必须在所有包的上一级。下面以 Hello World 代码为例，代码结构如图 1-12 所示。

Spring Boot 启动类在 com.zfx 包下，Spring Boot 控制层在 com.zfx.controller 包下，所以启动类能够扫描控制层相关的代码。包括控制层在内的所有 Spring 相关需要扫描并注册为 Bean 的代码都需要在 com.zfx.XXX 下，否则会报相关错误。若非该结构，则在代码中进行配置，让 Spring Boot 去扫描其他相关位置。

图 1-12

1.4.3　Spring Boot 扫描其他包下文件

Spring Boot 默认扫描当前包下级别的包与文件,若希望 Spring Boot 启动类可扫描其他包下文件，可利用@ComponentScan 注解或@Import 注解实现。

1. @ComponentScan 注解

为了让 Spring Boot 启动类扫描其他包下文件，可采用下述代码中@ComponentScan 注解的方案。在下述代码中并没有使用@SpringBootApplication 注解，但使用了其主要集成的@ComponentScan、@Configuration、@EnableAutoConfiguration 三个相关注解，由@Component-Scan 注解扫描，相关注册 Bean 仍生效。

```
//包扫描
@ComponentScan("com.zhang")
//组件扫描
@Configuration
//配置控制
@EnableAutoConfiguration
public class ApplicationMain {
    public static void main(String[] args) {
        SpringApplication.run(ApplicationMain.class, args);
    }
}
```

2. @Import 注解

为了让 Spring Boot 启动类扫描其他包下文件，可采用下述代码中@Import 注解的方案。在下述代码中使用了@Import 注解，用来引入其他未在 ApplicationMain 类以下层级的包相关可扫描的类。

```
@SpringBootApplication
@Import(SpringConfiguration.class)
public class ApplicationMain {//...}
```

图 1-13

以图 1-13 的工程结构为例，Spring Boot 的启动类 ApplicationMain 使用@SpringBoot- Application 注解扫描 bean、config、controller、dao 等包，但是无法扫描 com.zfx2.bean 和 com.zfx2.bean 包，所以需要使用其他相关注解扫描。

1.5 【实例】将端口号改成 9090

1.5.1 实例背景

Spring Boot 大大简化了所有 XML 相关的配置，但会集中到 application.properties 文件或 application.yml 文件中。

SSM 本身需要编写多个配置文件，如 applicationContext-server.xml、applicationContext-controller.xml，但事实上此类配置文件让初学者频繁出错，而有经验的程序员不断重复已经做好的部分，让整个项目变得特别臃肿。

而 Spring Boot 使用已经约定好的配置参数，完成相关配置，若无须更改 Spring Boot 事先写好的默认值参数，则不需要创建类似文件。

如果想修改默认的配置，如将 Spring Boot 启动后默认端口号 8080 更改为 9090，就需要新建配置文件 application.properties 或 application.yml，并在配置文件中加上一行代码 server.port=9090（该书写方式是在 application.properties 中的书写方式）。

另外，类似于 database 相关连接数据库等参数都需要写在 application.properties 资源配置文件中。application 是 Spring Boot 的一个设计核心，在日常工作中经常使用。

本实例将在 1.3 节 boot_01 工程的基础上，集成 application.properties 资源配置文件。

1.5.2　创建 application.properties 资源配置文件

如图 1-14 与图 1-15 所示，在 src/main/resource 下新建文件作为 Spring Boot 的资源配置文件，命名为 application.properties。

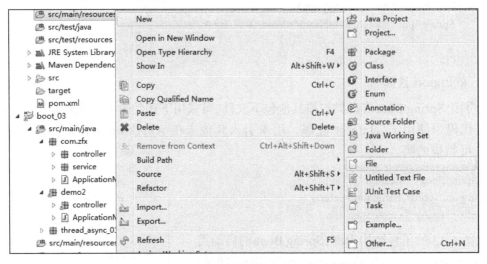

图 1-14

图 1-15

注意，此名是 Spring Boot 约定好的配置文件名，若有更改则 Spring Boot 无法扫描到配置文件。在前文中使用了 Spring Boot 的启动类注解@SpringBootApplication，其依靠@EnableAuto-Configuration、@SpringBootConfiguration 等相关注解来扫描默认地址下的 application.properties 资源配置文件，所以更改文件名后无法被正常扫描。

@EnableAutoConfiguration 注解：启动 Spring 应用程序上下文的自动配置，智能装备相关的 Spring Bean 内容。在装配的过程中，指定容器是否存在哪些 Bean、JVM 版本，指定路径是否存在指定文件等相关内容。在@EanbleAutoConfiguration 注解内部使用了@Import 注解，引入 EnableAutoConfigurationImportSelector.class 类的反射实例。

@Import 注解：@Import 是 Spring 的注解，通过导入的方式将实例加入 SpringIOC 容器中。

EnableAutoConfigurationImportSelector.class 实例：EnableAutoConfigurationImport-Selector.class 相当于@EnableAutoConfiguration 注解接口的实现，只有标记过@EnableAuto-Configuration 注解，才能使用 Spring Boot 相关功能。在 EnableAutoConfigurationImportSelector 实现类里含有 getAnnotationClass()函数，getAnnotationClass()函数将得到所有使用了@EnableAutoConfiguration 类的反射 Class<?>。此 Class<?>最主要的目的是整理整个工程的上下文，方便其他类调用，实现前文所说的@EnableAutoConfiguration 注解的作用。

1.5.3 增加资源配置文件中的配置信息

在 application.properties 资源配置文件中增加管理 Spring Boot 端口号的配置信息代码如下。

```
server.port=9090
```

1.5.4 运行结果

运行结果如图 1-16 所示，从下面日志看到此工程已通过 application.properties 资源配置文件把微服务的端口号从默认的 8080 改为 9090。Spring Boot 大部分配置信息都会存放到 application.properties 资源配置文件中。

```
2018-08-21  10:01:48.189   INFO 7416 --- [  restartedMain] s.b.c.e.t.TomcatEmbeddedServletContainer :
Tomcat started on port(s): 9090 (http)
```

图 1-16

1.5.5 实例易错点

1. 端口最大限制

端口的数值为无符号 short 类型，范围为 1～65535。在 Linux 系统中，1～1023 范围默认只有 Root 用户有权限使用，普通用户使用区间范围为 1025～65535，约 6 万。因此将端口设置为 99999 时将会报错，且无法启动 Spring Boot。端口最大限制异常报错信息如下。

```
***************************
APPLICATION FAILED TO START
***************************

Description:
The Tomcat connector configured to listen on port 99999 failed to start. The port may already be in use or the connector may be misconfigured.

Action:
Verify the connector's configuration, identify and stop any process that's listening on port 99999, or configure this application to listen on another port.
    2019-05-04 17:17:30.503  INFO 1976 --- [           main] ConfigServletWebServer-ApplicationContext : Closing org.springframework.boot.web.servlet.context.AnnotationConfigServletWebServerApplicationContext@4b5d6a01: startup date [Sat May 04 17:17:29 CST 2019]; root of context hierarchy
    2019-05-04 17:17:30.504  INFO 1976 --- [           main] o.s.j.e.a.Annotation- MBeanExporter       : Unregistering JMX-exposed beans on shutdown
```

2. 端口被占用

如果连续启动了两个端口为 9090 的 Spring Boot 程序，会出现启动报错且无法正常启动。

端口被占用异常报错信息如下。

```
***************************
APPLICATION FAILED TO START
***************************

Description:
The Tomcat connector configured to listen on port 9090 failed to start. The port may already be in use or the connector may be misconfigured.

Action:
Verify the connector's configuration, identify and stop any process that's listening on port 9090, or configure this application to listen on another port.
    2019-05-04 17:23:25.596  INFO 13100 --- [           main] ConfigServletWeb- ServerApplicationContext: Closing org.springframework.boot.web.servlet. context.AnnotationConfigServletWebServerApplicationContext@17046283: startup date [Sat May 04 17:23:24 CST 2019]; root of context hierarchy
    2019-05-04 17:23:25.597  INFO 13100 --- [           main] o.s.j.e.a.Annotation- MBeanExporter       : Unregistering JMX-exposed beans on shutdown
```

1.6 YAML 文件

YAML 文件的后缀为 yml，yml 文件和 xml、json、properties 文件一样，都是一种数据存储的文件，只是数据存储的格式不同。

1.6.1 YAML 文件简介

YAML 文件用纯文本的形式进行记录，直接用记事本进行书写，不需要编译。

Spring Boot 默认扫描资源配置文件，除.properties 文件外还读取 YAML 文件，即 yml 文件。Spring Boot 资源配置文件的约定命名分别为 application.properties、application.yml、bootstrap.properties、bootstrap.yml。

在代码中可以通过@Value、@ConfigurationProperties、@ApplicationContext 三种方式获取配置文件信息的形式，同时获得 YAML 文件内相关的配置信息且不需要更改任何代码。

1.6.2 YAML 文件的书写格式

（1）YAML 文件的 Key Value 结构应写为 Key: Value，冒号后若有值则要注意 Value 前必须有 1 个空格，若不写则无法正常响应文件。

（2）如果某个 Key 下有多个 Key，每个 Key 前就需要增加 2 个空格，如 org.zfx.boot.another.number，该 Key 的 number 前应该有 8 个空格。注意，不能使用 TAB 键缩进，这也是初学者的常见错误之一。

（3）每个 Key 结束后必须有一个"："作为结尾；如果没有，程序启动不会报错，但是用@Value 进行读取没有"："的值时，就会报空指针错误或 YAML 格式错误；如果有"："，就取空值。

（4）YAML 文件中没有"="，一切以"："为核心。

（5）YAML 文件的注释以"#"开头，但是如果写在"："后，必须在"："后加一个空格，否则报错，YAML 无法识别后会报 YAML 格式错误。

（6）不能通过@PropertySource 注解加载。

1.7 【实例】使用 YAML 配置文件

1.7.1 实例背景

本实例将使用 application.yml 资源配置文件替代 application.properties 资源配置文件，并对照 properties 文件编写存储相同数据的 YAML 文件。

1.7.2 原 properties 文件

在 application.properties 资源配置文件中输入其他格式的数据，目的是了解在 Spring Boot 的配置文件中，无论 YAML 文件还是 properties 文件，都能存储除 Spring Boot 预定义参数外的自定义参数，通过如@Value、@ConfigurationProperties、@EnableConfigurationProperties 等相关注解从资源配置文件中提取自定义参数。

在下述代码中 org.zfx.boot.date 能在 Java 代码中获取并转换成 List。org.zfx.boot.date 参数的配置也更能体现 YAML 文件和 properties 文件的书写区别。

原 application.properties 文件代码如下。

```
server.port=9090
another=zhangfangxing

org.zfx.boot.another.name=zhangfangxing
org.zfx.boot.another.number=1
org.zfx.boot.book=springboot
org.zfx.boot.date[0]=2015
org.zfx.boot.date[1]=2016
org.zfx.boot.date[2]=2017
org.zfx.boot.date[3]=2018
org.zfx.boot.date[4]=2019
```

1.7.3 转换格式后的 YAML 文件

将原 appliaction.properties 文件转换成 application.yml 文件后，application.yml 文件代码如下。在下述代码中，zfx 前有 2 个空格，boot 前有 4 个空格，another 前有 6 个空格，以此类推。

```
server:
  port: 9090
another: zhangfangxing

org:
  zfx:
    boot:
      another:
        name: zhangfangxing
        number: 1
      book: springboot
      date[0]: 2015
      date[1]: 2016
      date[2]: 2017
      date[3]: 2018
      date[4]: 2019
```

假设 org 是一个对象，则 org 对象含有 zfx 对象，zfx 对象含有 boot 对象，boot 对象含有 another 对象、book 对象、List<String>date（或 List<int>date、List<Date> date）对象。

1.7.4 实例易错点

1. YAML 文件不能用 TAB 键缩进

YAML 文件一定要用 2*n 空格来书写，若用 TAB 键则会报以下错误。

```
Caused by: org.yaml.snakeyaml.scanner.ScannerException: while scanning for the next token
found character '\t(TAB)' that cannot start any token. (Do not use \t(TAB) for indentation)
```

2. YAML 文件空格过多或过少

YAML 文件每个参数前一定要用 2*n 个空格来书写，若多一个空格或少一个空格则会报以下错误。

Caused by: org.yaml.snakeyaml.scanner.ScannerException: mapping values are not allowed here

1.8 【实例】通过单配置文件让工程适应多应用场景

1.8.1 实例背景

在生产中通常会运行 UAT（用户单元测试环境）或 SIT（用户验收测试集成环境）等多种环境，可能会造成频繁修改 application 资源配置文件的情况。

本实例在 boot_01 工程的基础上，基于 Profile 方式在一个资源配置文件下同时设置多个环境参数，每个环境所使用的参数不同。Spring Boot 微服务在启动时根据所输命令去适应应用场景。

1.8.2 更改 application.yml 文件

更改 application.yml 文件代码如下。

```yaml
server:
  port: 9999 #这里是注释
spring:
  profiles: uat
book:
  another: zhangfangxing
---
server:
  port: 9998
book:
  another: weiai
spring:
  profiles: sit
```

1.8.3 更改启动类

之前工程的启动类一直用 SpringApplication.run(ApplicationMain.class,args)启动，而下述代码的启动类使用了 SpringApplicationBuilder()，两者在运行上并无任何区别，只是 SpringApplication.run()不会额外设置 profile 文件等，大多采用默认状态。

更改 Spring Boot 的启动类 ApplicationMain.java 代码如下。

```java
package com.zfx;
import org.springframework.beans.factory.annotation.Autowired;
import org.springframework.boot.autoconfigure.SpringBootApplication;
```

```java
import org.springframework.boot.builder.SpringApplicationBuilder;
import org.springframework.context.ApplicationContext;
import org.springframework.web.bind.annotation.RequestMapping;
import org.springframework.web.bind.annotation.RestController;
@RestController
@SpringBootApplication
public class ApplicationMain {
    public static void main(String[] args) {
        new SpringApplicationBuilder(ApplicationMain.class).profiles(args[0]).run(args);
    }
    @Autowired
    private ApplicationContext context;
    @RequestMapping("/getName")
    public String getName() {
        return context.getEnvironment().getProperty("book.another");
    }
}
```

SpringApplication.run(.class,args)和 SpringApplicationBuilder().run(.class)在底层上都会进入 SpringApplication.java 类的 SpringApplication(ResourceLoader resourceLoader, Class<?>... sources)函数，加载配置文件相关信息，其底层代码如下。

```java
public SpringApplication(ResourceLoader resourceLoader, Class<?>... primarySources) {
    this.resourceLoader = resourceLoader;
    Assert.notNull(primarySources, "PrimarySources must not be null");
    this.primarySources = new LinkedHashSet<>(Arrays.asList(primarySources));
    this.webApplicationType = deduceWebApplicationType();
    setInitializers((Collection) getSpringFactoriesInstances(Application- ContextInitializer.class));
    setListeners((Collection) getSpringFactoriesInstances(Application- Listener.class));
    this.mainApplicationClass = deduceMainApplicationClass();
}
```

SpringApplication 构造函数相当于创建了一个新的 SpringApplication 实例，并且将加载应用程序上下文。之前做的更改实例配置等相关方式最终也是为了更改 SpringApplication 构造函数。该函数详解如下。

（1）ResourceLoader 参数：需要加载的资源与相关配置信息。

（2）primarySources 参数：额外引入的 Bean 资源。

（3）deduceWebApplicationType()函数：返回该 Web 程序的枚举类型，在 Spring Boot 中应用程序分为以下 3 种类型。

- 不需要在 Web 容器下运行，也是普通 Java 工程：WebApplicationType.NONE。
- 基于 Servlet 的 Web 应用：WebApplicationType.SERVLET。
- 响应式 reactive Web 应用：WebApplicationType.REACTIVE。响应式编程是 Spring5 推出的新特性，Spring 单独创建了一个 Spring WebFlux 框架，即响应式 Web 编程框架，其底层是由 Netty 框架提供的异步支持；它用少量的线程处理更高的并发，在支持异步方式的同时也支持同步方式，类似于WebSocket式的客户端与服务器双向监听。

（4）ApplicationListener监听接口：属于Spring Boot的核心组件之一，由多种方式实现，它基于观察者设计模式设计监听接口，用于应用程序的各种事件监听。

1.8.4 输入启动参数

在ApplicationMain.java类中右键单击工程，如图1-17所示，选择Run As→Run Configurations配置工程启动命令。因为在配置文件中编写了UAT和SIT两套资源环境，所以在Eclipse里需更改Run Configurations启动配置参数，否则Spring Boot无法判断用哪种方式启动。

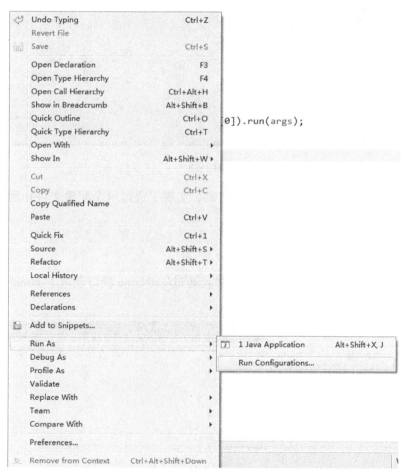

图1-17

Run Configurations对话框如图1-18所示，Run Configurations相当于java -jar命令行启动时后续输入的参数。在Program arguments文本框中书写UAT或SIT，便是将参数传入main方法的String[] args入参中，根据程序处理将工程配置成UAT或SIT相应的环境。若要配置JVM内存等相关信息，则在VM arguments文本框中书写。

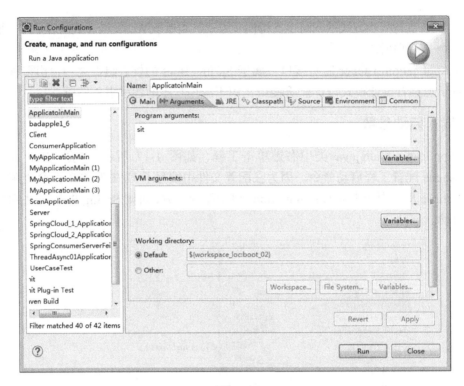

图 1-18

此时运行项目会通过 9998 端口进行启动,实现了通过单个配置文件让程序适应多个环境的目的。

1.8.5 运行结果

Spring Boot 已经用 9998 端口启动,并且通过/getName 接口获取了 book 下的 another 参数,运行结果如图 1-19 所示。

图 1-19

1.8.6 实例易错点

"---"(三个减号)用来分割多套配置信息。前面不可加空格,否则会报错,报错信息如下。

```
java.lang.IllegalStateException: Failed to load property source from location 'classpath:/application.yml'Caused
by: org.yaml.snakeyaml.parser.ParserException: while parsing a block mapping
    in 'reader', line 1, column 1:
        server:
```

（1）spring 下 profiles 后的 Value 值可自定义，即自定义某一区域内配置信息所起的名字，代码如下。

```
spring:
    profiles: sit2345
```

（2）spring 下 profiles 放置的位置只要在此区域内即可。

1.9 【实例】通过多配置文件使工程适应多应用场景

1.9.1 实例背景

1.8 节介绍了使用一个资源配置文件适应多个环境配置，也可使用多个资源配置文件适应多个环境配置。

本实例将创建 boot_02_10 工程，基于多个资源配置文件的方式同时设置多个环境参数，使整个工程项目结构在更加清晰的情况下适应多应用场景。

1.9.2 新建 SIT 和 UAT 环境所需资源配置文件

新建 boot_02 工程中 SIT 和 UAT 环境所需的资源配置文件 application-sit.properties 与 application-uat.properties 代码如下，以备后续适配所用。其中设置了不同的参数，待微服务的 application 资源配置文件对其中之一进行指向。

```
/boot_02/src/main/resources/application-sit.properties
server.port=9090
/boot_02/src/main/resources/application-uat.properties
server.port=8081
```

1.9.3 新建系统资源配置文件

编写 application.properties 资源配置文件代码如下，使用微服务的资源配置文件对 application-sit.properties 资源配置文件进行指向，将整个微服务工程适配到 SIT 环境中。

```
/boot_02/src/main/resources/application.properties
#指向所需的资源配置文件
#激活 SIT 环境所需的资源配置文件
spring.profiles.active=sit
```

1.9.4 编写启动类

使用常规的形式编写启动类，不需要增加其他改变。启动类 ApplicationMain.java 代码如下。

```
package com.zfx;
import org.springframework.boot.SpringApplication;
import org.springframework.boot.autoconfigure.SpringBootApplication;
@SpringBootApplication
public class ApplicationMain {
    public static void main(String[] args) {
        SpringApplication.run(ApplicationMain.class, args);
    }
}
```

1.9.5 当前项目结构

通过多配置文件使工程适应多应用场景的项目结构如图 1-20 所示，将多个 application 配置在 src/main/resources 文件夹中，通过 application.properties 文件对其进行引用即可。application-sit.properties 文件与 application-uat.properties 文件的前缀都是固定格式的，"-"后面的命名可自行定义。

1.9.6 运行结果

图 1-20

从以下运行日志中可以看到，系统已按照约定进行配置，自动指向了 application-sit.properties，获取其中信息，并按照 9090 端口进行启动，实现了通过多个配置文件使程序适应多个环境的能力，保证了 application 资源配置文件中的大部分参数都不需经常变动，并且提供给微服务适配多种环境的能力。

```
2018-08-23 12:46:19.746   INFO 11888 ---
[restartedMain] s.b.c.e.t.TomcatEmbeddedServletContainer : Tomcat started on port(s): 9090 (http)
```

1.10 微服务配置权重

Spring Boot 微服务分别使用了命令行、资源配置文件、Java 应用程序系统变量等方式对 Spring Boot 应用程序进行配置。若项目多次通过不同位置对端口号进行配置，Spring Boot 应用程序则会采用工程中权重最高的配置。本节将根据不同的配置信息类型、配置文件类型、配置文件存在位置介绍它们在微服务中所表现的权重效果。

1.10.1 资源配置信息类型的权重

不同配置方式的权重及输入方式如表 1-7 所示。

表 1-7 不同配置方式的权重及输入方式

权 重	输 入 方 式
1	命令行参数
2	通过 System.getProperties() 获取的 Java 系统参数

续表

权重	输入方式
3	操作系统环境变量
4	从 java:comp/env 得到的 JNDI 属性
5	通过 RandomValuePropertySource 生成的 random.*属性
6	应用程序外部的属性文件(通过 spring.config.location 参数)
7	应用程序内部的属性文件
8	在应用配置 Java 类（包含@Configuration 注解的 Java 类）中通过@PropertySource 注解声明的属性文件
9	通过 SpringApplication.setDefaultProperties 声明的默认属性

1.10.2 资源配置文件类型的权重

Spring Boot 默认扫描的配置文件分别为 application.properties、application.yml、bootstrap.properties、bootstrap.yml。若某个 Spring Boot 含有多个不同名称配置文件，不同名称的配置文件都对端口号进行配置，Spring Boot 应用程序则会采用权限最高的配置。

例如，同时含有 application.properties 配置文件和 bootstrap.yml 配置文件，两个文件都写了 server.port，将会以 application.properties 配置文件中的参数为准。不同配置文件的权重及输入方式如表 1-8 所示。

表 1-8 不同配置文件的权重及输入方式

权重	输入方式
1	application.properties
2	bootstrap.properties
3	application.yml
4	bootstrap.yml

1.10.3 资源配置文件存在位置与权重解读

Spring Boot 会在 4 个位置进行扫描，看看是否有相应的 application.properties 文件，即资源配置文件将放置在 4 个位置。Spring Boot 扫描不同位置资源配置文件的顺序如表 1-9 所示。

表 1-9 application 资源配置文件存在的位置及权重

权重	存在位置
1	src 下 config 文件夹(右键单击工程创建的 folder)
2	src 下根目录(右键单击工程直接创建 properties 文件)
3	src/main/resources 目录下 config 文件夹(在该目录下右键单击工程创建的 folder)
4	src/main/resources 目录下(在该目录下右键单击直接创建 properties 文件)

Spring Boot 的扫描机制是若某工程含有多个 application.properties 文件，则读完第 1 个

位置后不会读第 2 个位置或更后面的权重位置，即越靠前资源配置文件的权重越高。例如，读完第 1 个权重的配置文件，将 server.port 设置为 9090 后，第 2 个权重的配置文件的 server.port 为 19090，则以第 1 个权重为主，最后启动端口为 9090。

日常工作中的资源文件大多直接在 resources 文件夹下直接创建，这也是最合理的方式，工程中大多使用最低的权重来设置资源配置文件，方便应急时进行更改。

application 的位置与权重关系如图 1-21 所示。

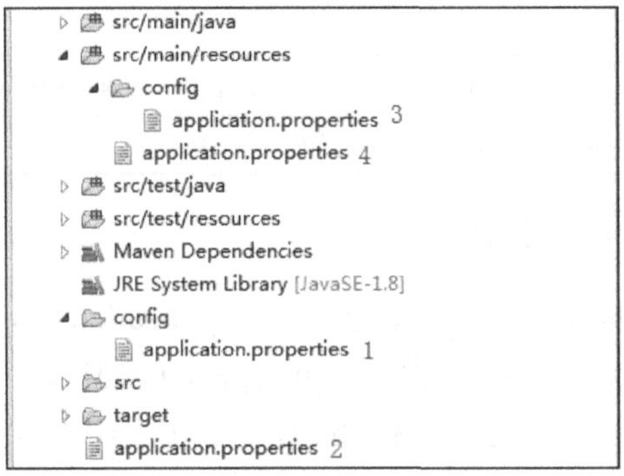

```
▷ 🗁 src/main/java
▲ 🗁 src/main/resources
   ▲ 🗁 config
       📄 application.properties  3
   📄 application.properties  4
▷ 🗁 src/test/java
▷ 🗁 src/test/resources
▷ 🖿 Maven Dependencies
▷ 🖿 JRE System Library [JavaSE-1.8]
▲ 🗁 config
   📄 application.properties  1
▷ 🗁 src
▷ 🗁 target
   📄 application.properties  2
```

图 1-21

1.11　本章小结

本章 1.3 节利用 Spring Boot 创建了第一个微服务应用程序。1.5 节将该应用程序的端口号修改为 9090。1.7 节将该应用程序的 properties 配置文件修改成了 YAML 配置文件。1.8 节通过单配置文件让工程适应多应用场景。1.9 节通过多配置文件使工程适应多环境进行开发。

在实际工程中，分布式的环境集成经常会使用多环境配置，多环境配置是微服务十分重要的部分。另外，在实际工程中建议多运用 bootstrap.yml 和 bootstrap.properties 文件，以方便日后维护。

1.12　习　　题

1. 创建 Spring Boot 工程，将实例端口号改为 50200，使接口参考 REST 风格返回相关数据。

2. 创建 Spring Boot 工程，使用 YAML 配置文件，在接口处使用注解获取配置文件相关信息，参考 REST 风格返回 YAML 配置文件中的数据，数据包括整数、浮点数、布尔、数组、集合、中文。同时，使 Spring Boot 工程根据启动入参的不同，返回的相关数据也不同，通过单个配置文件使工程适应多种应用场景。

3. 创建 Spring Boot 工程，使用多个 properties 配置文件，在接口处使用注解获取配置文件相关信息，参考 REST 风格返回 properties 配置文件中的数据，数据包括整数、浮点数、布尔、数组、集合、中文。同时，使 Spring Boot 工程根据 application.properties 文件指向的不同，返回的相关数据也不同，通过多个配置文件使工程适应多种应用场景。

第 2 章　分布式的注册中心

1.3 节创建了第一个微服务程序，从 1.1 节中 Dubbo+Zookeeper 分布式服务框架可以看出，无论分布式服务架构还是分布式微服务架构，都需要注册中心来管理众多服务/微服务。

当多个服务/微服务把自身的响应地址、服务内容等相关信息注册在注册中心上时，服务/微服务+注册中心所组成的架构就是分布式架构。

注册中心记录了服务和服务地址的映射关系。在分布式架构中，服务会注册到注册中心上，当服务需要调用其他服务时，在注册中心可找到其他服务的地址。分布式架构的基础是应用程序通过注册中心获得其他工程中提供服务的地址，并可以使用相关协议（如 Spring Cloud 的 HTTP 或 Dubbo 的 RPC）通过注册中心的地址，使当前工程享受其他工程提供的服务。

2.1　注册中心

分布式服务框架 Dubbo 用 Zookeeper 进行注册服务，负责提供互相调用的地址。ZooKeeper 是一个分布式的、开放源码的应用程序协调服务，是 Google Chubby 的一个开源实现，也是 Hadoop 和 Hbase 的重要组件。ZooKeeper 是一个为分布式应用提供一致性服务的软件，提供的功能包括配置维护、域名服务、分布式同步、组服务、分布式独享锁等。

只要两个服务注册在一个注册中心上，就已经可以看成初步的分布式。通过注册中心可以获得其他项目提供的服务，并从分布式获得其他项目目前是否良好运行的状态等信息，然后逐渐从注册中心辅助架构整套程序。注册中心相当于"通信录"，虽然无法直接拨打电话（调用服务），但会提供其他服务的地址和状态等相关信息。

2.1.1　Eureka 与 Consul 的区别

常见的服务注册中心有 Eureka、Consul、Zookeeper、Etcd、SmartStack、Serf、SkyDNS 等，下面详细介绍 Eureka 与 Consul 的区别。

1. Eureka

微服务分布式刚发布时，官方推荐使用 Spring Cloud Eureka 集成 Eureka 作为微服务的注册中心使用。Eureka 是 Netflix 公司开发的服务发现框架，其本身是一个基于 REST 的服务，Eureka 可以达到负载均衡和中间层服务故障转移的目的。Spring Cloud 将 Eureka 集成在其子项目 spring-cloud-netflix 中来实现 Spring Cloud 的服务发现功能。在应用启动后，将会向 Eureka Server 发送心跳，默认周期为 30s，如果 Eureka Server 在多个心跳周期内没有接收到某个节点的心跳，将会从服务注册表中把服务节点移除（默认 90s），Eureka 用心跳方式对项目进行健康检查，确保注册在 Eureka 注册中心上的服务都是健康可用的。但后来 Eureka 2.0 停产了。

虽然 Eureka 2.0 停产了，但 Eureka 1.0 的项目依旧被积极维护。Eureka 1.0 因为 Netflix 旗下其他项目还在使用，所以会一直更新下去。目前 Eureka 1.0 所提供的各种特殊功能虽然不多，但运行状况十分稳定。在 Eureka 2.0 停产后开始推荐使用 Consul 作为注册中心。注册中心 Eureka、ZooKeeper、Consul 并无较大差别。

Eureka 主要功能如下。

（1）Eureka 是一个服务发现工具。该体系结构主要是 Client/Server，通常每个客户端都使用嵌入式 SDK 进行注册和发现服务。

（2）Eureka 的健康检查主要依靠 HTTP 进行。服务在注册时有一个缓冲时间，在刚注册到 Eureka 时是无法使用的（虽然时间非常短暂）。

（3）Eureka 提供了一个弱一致性的服务方案。当客户端向服务器注册时，该服务器将尝试复制服务到其他服务器，但又不提供任何保证。可能有很短暂的一瞬间 Eureka 用过去版本的陈旧代码提供服务。弱一致性架构模式使 Eureka 具备高度可伸缩性的集群管理能力。

2. Consul

Consul 是一个服务网格（service mesh）解决方案，是一个分布式高可用的服务系统发现与配置工具、中间件，由 HashiCorp 公司研发。Consul 以 Go 语言编写，用 HTTP 方式对外提供服务，支持多数据中心，支持 K/V 键值对存储，可以构建完整的服务网格。

Consul 主要功能如下。

（1）服务发现：服务可以通过 HTTP 方式找到其他依赖 Consul 的服务。

（2）健康检查：检查服务是否正在正常响应。

（3）K/V 存储：提供了简易的 HTTP 接口满足用户的动态配置。

（4）安全业务通信：生成和分发服务的 TLS 证书，以建立相互的 TLS 连接，允许服务相互通信。

（5）多数据中心：支持多数据中心。不用构建额外的抽象层就能扩展到多个区域。

Consul 主要特点如下。

（1）Consul 提供了健康检查能力、K/V 存储和多数据中心意识。Consul 要求每个数据中心都有一组服务器，以及每个客户端上的一个代理，类似 Sideecar。Consul 代理允许大多数应用程序在不知情的情况下，通过配置文件执行服务注册，并通过 DNS 或负载均衡进行发现。

（2）Consul 提供了一致性保证，服务器使用 RAFT 协议，Consul 提供的健康检查能力包括 TCP、HTTP、Nagios/Sensu（网络监视工具）。

（3）Consul 的客户端节点依靠 Gossip Based 节点采样技术，即 Gossip 协议，通过采样技术进行健康检查。Consul 主动发现并发送请求给当前的 Consul Leader，即非集中式心跳，而集中性心跳将会成为可伸缩架构的障碍。

（4）Consul 提供了支持面向服务体系结构所需的工具包，包括服务发现、健康检查、服务锁定、K/V、多数据联合、事件系统和 ACL。Consul 试图将应用程序更改的内容降到最少，以避免通过 SDK 与本机进行集成。

（5）Consul 使用 Raft 算法保证一致性，比复杂的 Paxos 算法更直接。Zookeeper 采用 Paxos，而 Etcd 采用 Raft。

2.1.2　Consul 的相关术语

node：节点，需要 Consul 注册发现或配置管理的服务器。

agent：Consul 中的核心程序，以守护进程的方式在各节点运行，分为 client 和 server 两种启动模式。每个 agent 维护一套服务和注册发现及健康信息。

client：agent 以 client 模式启动的节点。在该模式下，节点会采集相关信息，通过 RPC 方式向 server 发送。

server：agent 以 server 模式启动的节点。一个数据中心至少包含 1 个 server 节点。也可以使用 3 个或 5 个 server 节点组建成集群，以保证高可用且不失效率。server 节点参与 Raft、维护成员信息、注册服务、健康检查等。

Consensus：共识协议，Consul 用其协商选出 leader。

Gossip：去中心化、容错并保证最终一致性的协议。因为 Consul 建立在 Serf 注册中心上，所以提供完整的 Gossip 协议。

members：成员，即对 Consul 成员的称呼。提供会员资格、故障检测和事件广播。

datacenter：数据中心，相当于整个 Consul 集群的名称，可以通过启动 Consul server 端来设置其名称。

2.1.3　Consul 的安装

下载 Consul 的二进制文件，将其放到 Linux 系统中进行解压缩后即可直接运行。在 Linux 系统中执行 consul 命令可看到以下内容输出，用来验证该文件是否正常可用。

```
$ consul
usage: Consul [--version] [--help] <command> [<args>]
Available commands are:
    agent           Runs a Consul agent
    event           Fire a new event
```

2.2　Consul 的常用命令

Consul 注册中心是 Linux 系统中可直接运行的二进制文件，本节介绍 Consul 注册中心的常用命令。

2.2.1 consul agent –dev

consul agent –dev 即开发模式启动，输入 consul -dev 可以在开发模式下启动 Consul。Dev 开发模式对于快速、方便地创建单节点 Consul 环境是非常有用的，但不能用于实际项目，因为 Dev 开发模式对任何状态信息都不能持久化。

consul agent -dev 的启动日志如下，Consul 已经启动并且输出了一些日志信息。其中，Node at 127.0.0.1:8300 [Leader] entering Leader state 信息是指在 Dev 开发模式下默认该 Consul 为主节点。

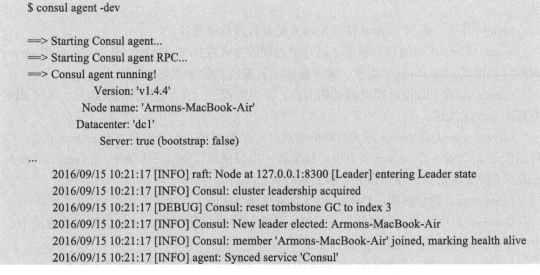

开发模式启动后，可以使用 http://127.0.0.1:8500/ui 访问 Consul 的 UI 页面，页面中提供了在 Consul 注册中心内注册的其他节点和服务信息。Consul 启动后的管理页面如图 2-1 所示。

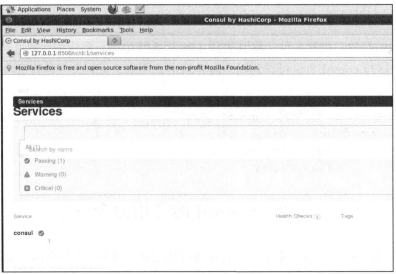

图 2-1

2.2.2 consul-members

consul-members 即列出成员信息,输入 consul -members 将会返回所有 Consul 节点中的成员信息。Dev 开发模式启动下只有一个节点信息。

如果要查询的 Consul 节点在其他服务器上,也可以输入命令 ./consul members -http-addr= 192.168.138.144:8500,-http-addr 可以指向到其他服务器上,集群中各节点状态的效果如下。

```
$ consul members

Node                   Address              Status   Type     Build   Protocol   DC
localhost.localdomain  172.20.20.11:8301    alive    server   1.4.4   2          dc1
```

此时输出了当前节点信息,包括 Consul 正在运行的地址、健康状态、在集群中的角色及版本信息。使用 HTTP API 给 Consul 发送请求时,也会返回相应的状态信息如下。

```
$ curl localhost:8500/v1/catalog/nodes

[
    {
        "ID": "a05fb778-b67d-f370-2f21-afd29f61a0b6",
        "Node": "localhost.localdomain",
        "Address": "127.0.0.1",
        "Datacenter": "dc1",
        "TaggedAddresses": {
            "lan": "127.0.0.1",
            "wan": "127.0.0.1"
        },
        "Meta": {
            "Consul-network-segment": ""
        },
        "CreateIndex": 9,
        "ModifyIndex": 10
    }
]
```

除 HTTP API 外,DNS 接口也可用来查询节点,在命令行中输入节点名称 localhost.localdomain,利用 members 命令查询信息如下。

```
$ dig @127.0.0.1 -p 8600 localhost.localdomain.node.consul

;; QUESTION SECTION:
;localhost.localdomain.node.Consul.  IN  A

;; ANSWER SECTION:
localhost.localdomain.node.Consul.  0  IN    A    127.0.0.1
```

```
;; ADDITIONAL SECTION:
localhost.localdomain.node.Consul. 0 IN     TXT     "Consul-network-segment="

;; Query time: 0 msec
;; SERVER: 127.0.0.1#8600(127.0.0.1)
;; WHEN: Sun May   5 05:48:41 2019
;; MSG SIZE   rcvd: 103
```

2.2.3　consul leave

consul leave 即离开集群，在当前 Consul 加入某一个 Consul 集群时需使用-join 命令；节点宕机后直接重启，也可以使用-rejoin 命令。

在当前 Consul 离开集群时需使用 consul leave 命令。consul leave 和 Ctrl+C 的区别在于，Ctrl+C 是"优雅"停止代理后，该 Consul 仍然会在集群中进行注册，集群中的 Server 服务器会不断提醒用户，该 Consul 无法正常使用，导致该 Consul 变成服务注册上的"僵尸"，但 consul leave 会取消其在集群中的注册。

2.2.4　agent 命令的常用配置参数

agent 命令代表运行一个 Consul agent 实例，如 consul agent -dev 命令。agent 命令的常用配置参数及释义如表 2-1 所示。

表 2-1　agent 命令的常用配置参数及释义

配 置 参 数	释　　义
-data -dir	指定 agent 存储状态的数据目录，是所有 agent 都必须使用的配置，对于 server 尤其重要，该目录在重新启动时是持久的；此外，目录必须支持文件系统锁定的使用，意味着某些类型的挂载文件夹（如 VirtualBox 共享文件夹）并不合适
-config-dir	指定 service 的配置文件和检查定义所在的位置，通常指定为"某一个路径/consul.d"（通常情况下，.d 表示一系列配置文件存放的目录）；Consul 将加载此目录中的所有文件，扩展名为.json 或.cl；加载顺序按字母顺序排列，可以多次指定此选项以加载多个目录，配置目录的子目录未加载
-config-file	指定一个要装载的配置文件，可以多次指定此选项以加载多个配置文件；若多次指定，稍后加载的配置文件将与先前加载的配置文件合并，后面的会合并前面的，相同的值覆盖
-config-format	要加载的配置文件的格式，通常 Consul 从.json 或.cl 扩展名检测配置文件的格式；若将此选项设置为 json 或 hcl，将强制 Consul 解释任何扩展名为.json 或.hcl 或没有扩展名的文件
-bootstrap	控制服务器是否处于 bootstrap 模式，bootstrap 模式下服务器可以自行选拔 Leader；注意，只需要有一个节点处于 bootstrap 模式下，若多个 bootstrap 模式下的节点都自选成 Leader，则无法保证 Consul 的一致性
-bootstrap-expect	通知 Consul server 现在准备加入的 server 节点个数，该参数为了延迟日志复制的启动，直到指定数量的 server 节点成功加入后才启动
-node	指定节点在集群中的名称，该名称在集群中必须是唯一的（默认采用机器的 host）
-bind	指明节点的 IP 地址，即内部集群通信应绑定的地址。-bind 是集群中所有其他节点都可以访问的 IP 地址，默认情况下为"0.0.0.0"；意味着 Consul 将绑定本地计算机上的所有地址

续表

配置参数	释义
-server	指定该节点为 server，在每个数据中心（DC）的 server 数推荐为 3 或 5（理想的是 3，最多不要超过 5）；所有的 server 都采用 Raft 一致性算法确保事务的一致性和线性化，事务修改了集群的状态，且集群的状态保存在每台 server 上以保证可用性
-client	将客户端接口（包括 HTTP 和 DNS 服务器）绑定的地址，默认情况下为"127.0.0.1"；指定节点为 client，若不指定节点为-server，则该节点是-client
-join	将节点加入集群
-dc	指定机器加入哪一个 DC 中
-advertise	advertise 地址用于更改向集群中的其他节点发布的地址
-datacenter	此标志控制代理正在运行的数据中心，若未提供，则默认为 DC1。Consul 支持多个数据中心，但依赖适当的配置；同一个数据中心中的节点应该位于单个 LAN 上
-domain	默认情况下，Consul 会响应 DNS 查询，此命令用于变更 DNS 域，该域中所有的查询都被 Consul 处理，并且不被递归解析
-enable-script-checks	将控制是否执行脚本的健康检查在此代理上启用，默认为 false，在 Consul 0.9.0 版本中添加

2.2.5　HTTP API

表 2-2 所列的是 Consul 中常见的 HTTP API。

表 2-2　Consul 中常见的 HTTP API

HTTP 地址	API 功能释义
ip:8500/v1/agent/members	列出所有的 members
ip:8500/v1/agent/checks	返回本地 agent 注册的所有检查(包括配置文件和 HTTP 接口)
ip:8500/v1/agent/reload	重新加载 agent
ip:8500/v1/agent/leave	"优雅"退出和关闭
ip:8500/v1/agent/force-leave/<node>>	强制退出和关闭
ip:8500/v1/agent/self	返回本地 agent 的配置和成员信息
ip:8500/v1/agent/services	返回本地 agent 注册的所有服务
ip:8500/v1/agent/join/<address>	触发本地 agent 加入 node
ip:8500/v1/agent/check/register	在本地 agent 中增加一个检查项，使用 PUT 方法传输 json 格式的一个数据
ip:8500/v1/agent/check/deregister/<checkID>	注销本地 agent 的一个检查项
ip:8500/v1/agent/check/pass/<checkID>	设置本地一个检查项的状态为 passing
ip:8500/v1/agent/check/warn/<checkID>	设置本地一个检查项的状态为 warning
ip:8500/v1/agent/check/fail/<checkID>	设置本地一个检查项的状态为 critical
ip:8500/v1/agent/service/register	在本地 agent 中增加一个新的服务项，使用 PUT 方法传输 json 格式的一个数据
ip:8500/v1/agent/service/deregister/<serviceID>	注销本地 agent 的一个服务项

2.3 【实例】创建第一个微服务分布式项目

2.3.1 实例背景

2.2 节简要介绍了 Consul 的部分基本操作，本实例将创建 cloud_01 工程，并以 Consul 集群搭建/部署过程的形式详解如何组建 Consul 集群。通过 Spring Cloud Consul 框架整合 Consul 注册中心集群，最后在 Consul 注册中心管理页面中查看相关的 Spring Boot 微服务工程。

2.3.2 搭建 Consul 集群

Consul 集群是指在相同的 Consul 数据中心中，同时运行多台 Consul 注册中心，让 Consul 共享存储的微服务信息，保证高并发下的注册中心运行效果。

1．准备服务器

本实例将使用 4 台虚拟服务器进行搭建，准备的服务器如表 2-3 所示。

表 2-3 为了搭建 Consul 集群准备的虚拟服务器列表

IP	Type	备注
192.168.138.144	Consul 的 Server 端	Linux 服务器
192.168.138.140	Consul 的 Client 端	Linux 服务器
192.168.138.147	Consul 的 Client 端	Linux 服务器
192.168.138.1	Consul 的 Client 端	Windows 服务器

2．下载及部署

将 Consul 应用程序分别传输到不同的服务器上，如图 2-2（Windows）和图 2-3（Linux）所示。

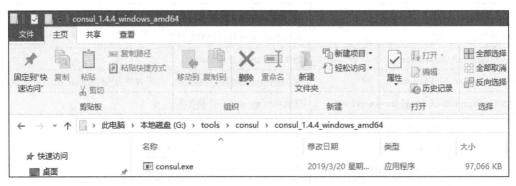

图 2-2

在 Linux 系统中使用 unzip 命令可将 ZIP 文件解压，解压之后即可直接使用 ./consul 命令启动。其他启动 Consul 的方式如下。

（1）更改/etc/profile 文件，更改相应的环境变量。
（2）将 Consul 文件传输到/usr/local/bin 文件夹中。
（3）配置快捷启动。
（4）编写启动脚本。

图 2-3

3．逐个启动 Consul 注册中心

192.168.138.144 的 Server 端 Consul 节点启动命令如下。

./consul agent -server -bootstrap-expect 1 -data-dir=/data/consul -node=consul-server-144-bind= 192.168.138.144 -client=0.0.0.0 -ui&

其他 Consul 节点只需依次更改命令中的-bind 参数为当前需要配置的 IP 地址，即可启动其他 Consul 节点。192.168.138.140/192.168.138.147 等 Linux 系统的 Client 端 Consul 节点启动命令如下。

./consul agent -server -bootstrap-expect 1 -data-dir=/data/consul -node=consul-server-140 -bind= 192.168.138.140 -client=0.0.0.0 -ui -join 192.168.138.144&

在-server 生产级别启动注册中心，至少含有 3 台-server 服务器。若只进行测试，则可以只有一台-server 服务器，与-dev 命令效果相同，如图 2-4 所示。

图 2-4

此时已经成功搭建了一个 Consul 集群。每个 Consul 都可以给 n 个微服务进行注册，视项目大小可酌情添加 Consul 的 Client 节点，但不宜过多，以免管理维护困难。另外，也可以把 node 参数中的 Value 都换成 Consul 所在服务器的 IP，也容易区分各 Consul 节点。

4．利用配置文件启动注册中心并修改端口号

如果在启动 192.168.138.147 服务器的注册中心时端口被占用，可以通过命令修改端口号并进行启动。如果命令过长不方便输入，可以编写一个 json 配置文件，在 json 配置文件中输入端口号等相关配置信息，一次性更改所有配置。json 配置文件代码如下。

```
{
  "ports":{
    "server":9300,
    "serf_lan":9301,
    "serf_wan":9302,
    "http":9500,
    "dns":9600
  },
  "data_dir": "/data/Consul"
}
```

修改 json 配置文件可以更改配置信息，并且相关命令参数也都可以写入文件中。Consul 注册中心的默认端口号如表 2-4 所示。

表 2-4　Consul 注册中心的默认端口号

服务与协议	端　口　号
dns	8600
http	8500
serf_lan	8301
serf_wan	8302
server	8300

在扫描配置文件时，通过 -config-file 命令可以扫描配置文件。

5．集群搭建后的效果

当前集群下的 Server 端 UI 展示效果如图 2-5 所示。

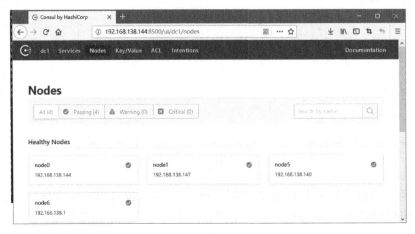

图 2-5

当前集群下的命令行展示效果如图 2-6 所示。

```
[root@localhost consul]# ./consul members -http-addr=192.168.138.144:8500
Node    Address              Status  Type    Build  Protocol  DC   Segment
node0   192.168.138.144:8301 alive   server  1.4.4  2         dc1  <all>
node1   192.168.138.147:8301 alive   client  1.4.4  2         dc1  <default>
node5   192.168.138.140:8301 alive   client  1.4.4  2         dc1  <default>
node6   192.168.138.1:8301   alive   client  1.4.4  2         dc1  <default>
```

图 2-6

2.3.3 创建微服务工程编写相应依赖文件

搭建 Consul 集群后，需将 Spring Boot 微服务工程注册到 Consul 注册中心上，使 Consul 注册中心获得微服务工程的相关信息，微服务依赖 Spring Boot 微服务相关的文件、Consul 整合文件等，pom.xml 文件依赖代码如下。

```xml
<project xmlns="http://maven.apache.org/POM/4.0.0" xmlns:xsi="http://www.w3.org/2001/XMLSchema-instance" xsi:schemaLocation="http://maven.apache.org/POM/4.0.0 http://maven.apache.org/xsd/maven-4.0.0.xsd">
    <modelVersion>4.0.0</modelVersion>
    <groupId>org.zfx.cloud</groupId>
    <artifactId>cloud01</artifactId>
    <version>0.0.1-SNAPSHOT</version>
    <parent>
        <groupId>org.springframework.boot</groupId>
        <artifactId>spring-boot-starter-parent</artifactId>
        <version>2.0.4.RELEASE</version>
        <relativePath />
    </parent>
    <properties>
        <project.build.sourceEncoding>UTF-8</project.build.sourceEncoding>
        <project.reporting.outputEncoding>UTF-8</project.reporting.outputEncoding>
        <java.version>1.8</java.version>
        <spring-cloud.version>Finchley.RELEASE</spring-cloud.version>
    </properties>
    <dependencies>
        <dependency>
            <groupId>org.springframework.cloud</groupId>
            <artifactId>spring-cloud-starter-consul-discovery</artifactId>
        </dependency>
        <dependency>
            <groupId>org.springframework.boot</groupId>
            <artifactId>spring-boot-starter-web</artifactId>
        </dependency>
    </dependencies>
    <dependencyManagement>
        <dependencies>
            <dependency>
```

```xml
            <groupId>org.springframework.cloud</groupId>
            <artifactId>spring-cloud-dependencies</artifactId>
            <version>${spring-cloud.version}</version>
            <type>pom</type>
            <scope>import</scope>
        </dependency>
    </dependencies>
</dependencyManagement>
<build>
    <plugins>
        <plugin>
            <groupId>org.springframework.boot</groupId>
            <artifactId>spring-boot-maven-plugin</artifactId>
        </plugin>
    </plugins>
</build>
</project>
```

  本书中 Spring Cloud 版本是 Finchley.RELEASE，Spring Cloud 版本的名字都用英文命名，因为 Spring Cloud 每个版本中会集成 Spring Boot 的相关插件和 Spring Cloud 的相关组件。

  而 Spring Cloud 因为集成的插件、组件过多，如 Spring Cloud Config，Spring Cloud Netflix，Spring Cloud Bus，所以经常发生版本不兼容的情况。

2.3.4 Spring Cloud 和 Spring Boot 的版本对应关系

  Spring Cloud 与 Spring Boot 的版本对应关系如表 2-5 所示，如果 Spring Cloud 与 Spring Boot 不兼容，会导致程序出现大量未知异常。

  Spring Boot 2.0 以下版本，功能较少。如 Spring Data MongoDB，在 Spring Boot 1.5 版本只提供 MongoRepository 仓库映射的方式进行 CURD，在 Spring Boot 2.0 版本更新了 MongoTemplate 模板，以方便管理 MongoDB 数据库。

表 2-5　Spring Cloud 与 Spring Boot 的版本对应关系

Spring Cloud	Spring Boot
Finchley	兼容 Spring Boot 2.0，不兼容 Spring Boot 1.5
Dalston、Edgware	兼容 Spring Boot 1.5，不兼容 Spring Boot 2.0
Camden	兼容 Spring Boot 1.4 和 Spring Boot 1.5
Brixton	兼容 Spring Boot 1.3 和 Spring Boot 1.4
Angel	兼容 Spring Boot 1.2

2.3.5 编写微服务 YAML 资源配置文件

  Spring Cloud Consul 初步整合 Consul 注册中心的 application.yml 资源配置文件，代码如下。

```yaml
spring:
  cloud:
    consul:
      host: 192.168.138.144
      port: 8500
      discovery:
        healthCheckPath: /health
        healthCheckInterval: 15s
        instance-id: cloud01-id
        hostname: 192.168.138.1
  application:
    name: cloud01-application
server:
  port: 50123
```

spring.cloud.consul.host 和 spring.cloud.consul.port 代表要去注册的注册中心地址，而 spring.cloud.consul.discovery.hostname 代表当前应用程序运行的地址。

spring.application.name 是 Spring Boot 微服务设定的名称，在 Consul 的 Web UI 管理页面中也会显示微服务工程的名字。

spring.cloud.consul.discovery.instance-id 是在 Consul 中注册的名字。instanceId 可省略，Consul 将显示 ${spring.application.name}。注意，spring.cloud.consul.discovery.instance-id 的名字必须唯一，默认情况下 Consul 使用 SpringApplicationContext ID 中的 ID 进行注册。Spring 应用程序上下文 ID 为 ${spring.application.name}:comma,separated,profiles:${server.port}，代码如下。

```yaml
spring:
  cloud:
    consul:
      discovery:
        instanceId: ${spring.application.name}:${vcap.application.instance_id:${spring.application.instance_id:${random.value}}}
```

spring.consul.discovery.healthCheckPath 为 Consul 注册中心检查应用程序是否健康的接口路径，默认路径为 /health。该路径必须返回 200，否则 Consul 会认为微服务处于无法正常运行的状态。为避免应用程序接口重复，也可以写成如下代码。

```
healthCheckPath: ${management.server.servlet.context-path}/health
```

spring.cloud.discovery.healthCheckInterval 为 Consul 注册中心检查应用程序是否健康的间隔时间，可设置为 1 分钟。

在 YAML 或 properties 文件中书写 ${} 相关内容都是 Spring Boot 整合 SpringEL 所提供的 EL 表达式能力，可以获取 Spring 容器中的相关参数。例如，在 YAML 文件中编写 port: ${server.port}，代表从 Spring 容器的上下文中获得了 server.port 参数，并以此赋值给 port 参数。

2.3.6 编写微服务启动类注册到 Consul 上

Spring Cloud 集成 Eureka 时使用@EnableEurekaClient 注解连接 Eureka 注册中心，而集成 Consul 或 Zookeeper 等其他注册中心时使用 @EnableDiscoveryClient 注解。@EnableEurekaClient 注解与@EnableDiscoveryClient 注解的用法基本相同，@EnableDiscoveryClient 也可以用于注册到 Eureka 上。@EnableDiscoveryClient 注解基于 spring-cloud-commons 包，@EnableEurekaClient 注解基于 spring-cloud-netflix 包。

Spring Boot 微服务整合 Consul 创建微服务分布式项目的启动类 ApplicationMain.java，代码如下。

```java
package com.zfx;
import org.springframework.boot.SpringApplication;
import org.springframework.boot.autoconfigure.SpringBootApplication;
import org.springframework.cloud.client.discovery.EnableDiscoveryClient;
import org.springframework.web.bind.annotation.RequestMapping;
import org.springframework.web.bind.annotation.RestController;
@RestController
@EnableDiscoveryClient
@SpringBootApplication
public class ApplicationMain {
   @RequestMapping("/health")
   public String health() {
        return "200";//此应用程序将给 Consul 返回 200，告诉 Consul 此应用程序还在正常运行
   }
   public static void main(String[] args) {
        SpringApplication.run(ApplicationMain.class, args);
   }
}
```

如果不编写 health()函数，Consul 注册中心的 Web UI 界面会报错，告诉用户这个微服务无法正常连接。health() 函数在 YAML 资源配置文件中配置，告诉 Consul 注册中心调用该函数检测微服务是否正在正常运行。

2.3.7 当前项目结构

如图 2-7 所示，利用 Spring Cloud 集成 Consul 注册中心，只需要一个启动类和一个资源配置文件即可，除健康检查 Health()接口和@EnableDiscoveryClient 启动注册中心接口外，不需要其他任何 Java 代码。

图 2-7

2.3.8 运行结果

启动 Spring Boot 应用程序后，可以在 Consul 的 UI 界面中观察应用程序，如图 2-8 所示。

图 2-8

Serf Health Status：表示当前 Consul 节点正在良好地运行（或其他状态）。

Service'cloud01-application' check：其下方文字表示 cloud01-application 的健康检查正在良好运行。

如果复制 cloud-01 工程，并更改其端口号和工程名称等相关参数，再次注册到 Consul 中的效果如图 2-9 所示。

图 2-9

从图中可以看到，两个应用程序都在良好运行，此时第一个分布式应用程序已经搭建成功。除了在 192.168.138.144 的 Consul 服务器上可以看到相关信息，在其他服务器上也可以看到相关信息。在 Consul 的 server 集群中，每个 server 的数据都是共享的，如图 2-10 和图 2-11 所示。

图 2-10

图 2-11

2.3.9 实例易错点

1. Consul 无法正常检测到健康接口

如图 2-12 所示，Consul 检测到 cloud033-application 应用程序的健康接口失效，报错时只能检查资源配置文件中的 spring.consul.discovery.healthCheckPath 健康接口路径编写是否正确，并且是否正常返回 200。

图 2-12

2. 实例化失败

如果含有 creating bean 等相关报错信息，通常都由配置信息不全所产生。可优先尝试用静态书写相关内容，确认无误后再更改为配置文件的动态形态。

注意，不要遗忘参数 spring.application.name。spring.application.name 是 Spring Boot 对当前应用程序命名的参数，Consul 也以此开始对整个应用程序进行实例化；是分布式最重要的参数之一。实例化失败报错信息如下。

Error creating bean with name 'org.springframework.cloud.client.serviceregistry.AutoServiceRegistration-AutoConfiguration': Unsatisfied dependency expressed through field 'autoServiceRegistration';

3. @SpringCloudApplication 注解报错

注意，启动程序的启动类使用的依旧是 @SpringBootApplication 注解，若写成 @SpringCloudApplication 注解，则会报错。

Spring Cloud 认为每个微服务分布式的应用程序，都应含有 Spring Boot 相关启动注解、注册中心相关注解和熔断器注解，并以此推出@SpringCloudApplication 注解整合分布式上述 3 个注解。但是应用程序目前没有集成其他分布式框架，所以使用@SpringCloudApplication 注解会让 Spring Cloud 检测不到一些应该编写的配置文件，从而引发报错，错误信息如下。

Error processing condition on org.springframework.boot.autoconfigure.context.PropertyPlaceholderAuto-Configuration.propertySourcesPlaceholderConfigurer

2.4 【实例】通过代码获取 Consul 中的服务信息

2.4.1 实例背景

大部分注册中心底层实现原理都基本相同，主要分为注册中心、服务提供者、服务消费者 3 个角色。

注册中心会在自身创建相关的服务注册表，让多个服务提供者在表上注册，组织服务提供者形成集群，让服务消费者通过服务注册表获取具体的服务访问地址，访问地址通常以 IP+端口号的 HTTP 形式在 Consul 内进行存储。在服务消费者调用相关地址后，可以获得服务提供者的功能。在 Consul 的 Web UI 控制器中 Consul 自身的节点为 node，每个不同的 service 会注册到不同的 node 中，被注册的 node 将会存储一系列 service 的相关信息，如该 service 的 IP、端口、调用方式、协议、序列化方式等。从此时开始，对于 node 来说

service 是被注册的 node 中的一个节点。无论一个 node 中有多少 service，对于该 node 来说都是整合了众多不同地址、不同命名的节点。node 之间会不断相互通信，发送到 node 中的 service 节点上，就完成了分布式的思想，也是最基础的注册中心能力。在此基础上，每个注册中心都会在保证自身稳定性、性能、文件大小的情况下做出不同的功能（如 Consul 中的 K/V 和 ACL），提供不同的特性（如 Consul 中替代心跳的 Gossip 协议），用户可根据不同的需求场景选择不同的注册中心。

本实例将更改 2.3 节中的 cloud_01 工程，并使用 Spring Boot 应用程序通过 Consul 注册中心，获得在 Consul 注册中心上注册的其他微服务及相关地址等信息。

2.4.2 编写获得其他注册服务的代码

为方便测试，编写 ConsulController.java 类中接口，返回注册中心中相关服务的数据，代码如下。在 ConsulController 接口中通过 Spring Cloud 的 LoadBalancerClient 接口和 DiscoveryClient 接口，获得 Consul 注册中心中其他注册服务的相关地址。

```java
package com.zfx.controller;
import org.springframework.beans.factory.annotation.Autowired;
import org.springframework.cloud.client.discovery.DiscoveryClient;
import org.springframework.cloud.client.loadbalancer.LoadBalancerClient;
import org.springframework.web.bind.annotation.PathVariable;
import org.springframework.web.bind.annotation.RequestMapping;
import org.springframework.web.bind.annotation.RestController;
@RestController
public class ConsulController {
    @Autowired
    private LoadBalancerClient loadBalancer;
    @Autowired
    private DiscoveryClient discoveryClient;
    /**
     * 从所有服务中获得一个服务的地址
     **/
    @RequestMapping("/discover/{serviceName}")
    public Object discover(@PathVariable String serviceName) {
        return loadBalancer.choose(serviceName).getUri().toString();
    }
    /**
     * 获取某个服务的全部信息，包括僵尸节点和负载节点
     */
    @RequestMapping("/services/{serviceName}")
    public Object getServicesBy(@PathVariable String serviceName) {
        return discoveryClient.getInstances(serviceName);
    }
}
```

LoadBalancerClient 类：负载均衡器，继承自 ServiceInstanceChooser 接口。负载均衡是指分摊到多个操作单元上进行执行，如 Web 服务器、FTP 服务器、企业关键应用服务器和其他关键任务服务器等，共同完成工作任务。LoadBalancerClient 负载均衡器含有两个抽象函数 execute、reconstructURI，其中 execute 被重载一次，用来执行（来自）负载均衡器的 Service 实例指定服务的请求；reconstructURI 用来为 Service 创建一个合适的 URI，入参为 ServiceInstance 对象。

ServiceInstance 对象：Spring Cloud 整理注册 Service 服务实例的 Java Bean，包括注册 Service 服务实例的地址、端口号、服务 ID 等相关信息。

ServiceInstanceChooser 接口：只含有一个待实现的 choose(serviceID)抽象函数，由负载均衡器指定某个服务 ID 获得注册在 Consul 注册中心上的某个 Service 实例。

DiscoveryClient 接口：发现服务（如 Eureka 注册中心或 Consul 注册中心上的已注册服务），通常用于读取注册服务的相关信息。DiscoveryClient 接口中含有 3 个待实现抽象函数，分别为 description()返回健康检查状态函数，getInstances(String ServiceID)根据服务 ID 获得 Service 服务实例信息函数，getServices()获得全部已注册的服务 ID。

2.4.3 运行结果

如图 2-13 所示，通过 Java 代码获取 Consul 注册中心上其他服务的注册服务信息，接口返回 json 数据格式的信息。

可使用 HttpClient 等相关工具，通过 Consul 注册中心上注册的其他服务 URL 等相关信息调用其他服务的接口。例如，图 2-13 中 cloud02-application 微服务的 IP 为 localhost:50124，只要知道 cloud02-application 微服务提供接口的名称及入参，即可进行调用。

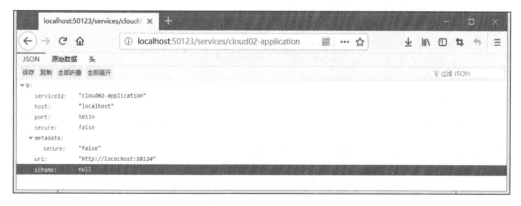

图 2-13

2.4.4 实例易错点

不可忽视服务注册 Consul 的初始化时间，如果在上述代码中并没有代码上的 Bug，可在调用相关接口时总返回空指针异常等相关错误，就很有可能是启动服务和调用接口的操作时间过快导致的。例如，心跳配置的是 15s，则初始化时间应该是在 15s 以上，15s 以后才能根据接口返回其他服务相关信息。错误通常如下。

```
java.lang.NullPointerException: null
```

2.5 【实例】Spring Cloud 操作 Consul 的 K/V 存储

2.5.1 实例背景

K/V 存储是 Consul 的特点之一。与 Redis 缓存不同的是，Consul K/V 存储的功能是辅助分布式环境下每个微服务的配置等。

一旦微服务过多，每个微服务都需要大量的配置信息，这样很容易造成维护上的困难。在分布式刚兴起的阶段，Spring Cloud 给出了 Spring Cloud Config 组件作为解决方案，其原理是把配置文件集中在 Git 或 SVN 等相关版本控制软件中，方便进行统一维护管理，然后在应用程序中使用 Spring Cloud Config 组件读取和调用配置。因为 Consul 自带 K/V 存储，所以已经基本替代了 Spring Cloud Config，不但减少了学习成本，更减少了 Git 和 SVN 的管理控制成本。如果用 SVN 管理配置文件，就需要给 SVN 做高可用和容灾处理，以免无法读取配置文件从而不能正常运行项目。

本实例继续更改 cloud-01 工程，首先通过 Consul 的 Web UI 存储一部分参数，然后通过 cloud-01 工程对 Consul 的 K/V 存储上的数据进行读取操作。

2.5.2 添加依赖

pom.xml 文件所需增加代码如下。若用<parent>继承 Spring Boot 的方式，则不需要再写<version>版本号，Spring Boot 会统一管理。

```xml
<dependency>
    <groupId>org.springframework.cloud</groupId>
    <artifactId>spring-cloud-starter-consul-config</artifactId>
    <version>2.0.0.RELEASE</version>
</dependency>
```

注意，操作 Consul 的 K/V 存储的主要 API 都在 spring-cloud-starter-consul-config 包中。

2.5.3 利用 Consul 的 UI 界面添加 K/V 存储

Consul 的 UI 界面进行数据存储有 3 种格式，分别为 json、yml 和 hcl。

如图 2-14 所示，在 Key or folder 文本框中输入相关的 Key，在 Value 文本框中输入相应内容，右下角选择定义 Value 的数据格式，右上角 Code 表示是否对 Value 进行数据格式校验。整个界面可看成一个小型的 IDE 程序。

图 2-14 中的 Key 值为 config/cloud01,dev/data。

config：spring cloud consul config 的默认存放地址（文件夹），也可通过 bootstrap 配置文件更改 spring.cloud.consul.config.prefix 参数。

cloud01：分布式该微服务的名称，即${spring.application.name}值，该名称不同于 project

工程名，不同于 URL 中追加的工程映射的路径名，而是自定义独立存在的微服务名。

dev：cloud01 应用程序的 profile 环境规定，即${spring.profiles.active}值，与微服务名连起来写为 cloud01,dev。

data：spring cloud consul config 默认存放的 Key 名，也可通过 bootstrap 配置文件更改 spring.cloud.consul.config.data-key 参数并进行自定义。

Value 内部需要书写对应 config/cloud01,dev/data 该 Key 值的 Value。

Value 书写 YAML 格式，其中 jdbc 是 Value 的头部标签，将按照该标签进行取值，使程序可以映射到资源配置中。jdbc.my.year 展示了在次级目录下如何获取相关参数。

图 2-14

Value 内部代码如下，其格式与 YAML 文件保持一致。

```
jdbc:
  username: myusername
  password: mypassword
  my:
    year: 2019
```

2.5.4　编写 YAML 资源配置文件对应 K/V 存储

YAML 资源配置文件将与 2.5.3 节中存储的相关数据进行对应。application.yml 资源配置文件代码如下。

```
spring:
  profiles:
    active: dev
  application:
    name: cloud01
  cloud:
```

```yaml
    consul:
      discovery:
        hostname: localhost
        port: 8500
      config:
        fail-fast: true
        format: YAML
        prefix: config
        defaultContext: application
        profileSeparator: ','
        data-key: data
```

spring.cloud.consul.config.fail-fast：true 代表开启快速失败。快速失败是指在 Java 用迭代器遍历一个集合对象时，如果遍历过程中对集合对象的内容进行了修改（如增加、删除、修改），就会抛出 Concurrent Modification Exception 异常。开启快速失败代表应用程序要求接收到的数据是最新的，如果可能出现脏读等情况，宁愿触发失败也不接收数据。快速失败的原理是迭代器遍历集合中的内容，并且遍历前定义一个变量，若在遍历期间内容发生变化，则会改变变量的值。每当迭代器.hasNext()/next()遍历下一个元素前，都会检测定义的变量是否正常，否则抛出相关异常。

spring.cloud.consul.config.format: YAML 是 Consul 存储中定义的格式。

spring.cloud.consul.config.prefix: config 是基础文件夹名称，对应 Key 键下 config/cloud01,dev/data 的 config 文件夹名称。

spring.cloud.consul.config.defaultContext: application 是所有应用程序默认文件夹，若有 Key 值，则直接进行读取。

spring.cloud.consul.profileSeparator: ','用于设置分隔符的值，使用配置文件在属性源中分隔配置文件名称。

spring.cloud.consul.data-key: data 是 Key 值名称对应 Key 键下 config/cloud01,dev/data 的 data 值。

Key 值 config/cloud01,dev/data 中的 cloud01 对应 spring.application.name，dev 对应 spring.profiles.active。

2.5.5 编写 MyConfig.java 文件对应相关 K/V 存储

编写 MyConfig.java 实体类，保证 Java 程序与 2.5.3 节中 Consul K/V 存储的数据相互对应，其代码如下。

```java
package com.zfx.config;
import org.springframework.boot.context.properties.ConfigurationProperties;
@RefreshScope
@ConfigurationProperties(prefix = "jdbc")
public class MyConfig {
    private String username;
    private String password;
```

```
    private My my;
    //省略 setter/getter，My 对象里只有一个 private String year 参数和 setter/getter 函数
}
```

@ConfigurationProperties 注解对应 Consul 中 config/cloud01,dev/data 的 Key 值下的 YAML 头部，即获得 JDBC 头部以下的参数。如果没有头部 YAML 文件，@ConfigurationProperties 注解内部设置为空即可。注意，此时已经截掉了 JDBC，JDBC 剩下的 username 和 password 必须与 Consul K/V 中的名字对应，否则 Spring Cloud 无法进行匹配和映射。

@RefreshScope 注解的作用是监控 Consul 的 K/V 存储刷新，一旦 Consul 上的数据被更改，获取的参数也会相应被更改。

2.5.6 调用 MyConfig.java 中的参数

为方便测试，创建 ConfigController.java 接口类，在接口中返回 Consul K/V 存储内的相关参数，代码如下。将 MyConfig 类通过@Autowired 注解注入 ConfigController 类后，可直接获得 Consul K/V 存储内的相关参数。

```
package com.zfx.controller;
import org.springframework.beans.factory.annotation.Autowired;
import org.springframework.web.bind.annotation.RequestMapping;
import org.springframework.web.bind.annotation.RestController;
import com.zfx.config.MyConfig;
@RestController
public class ConfigController {
    @Autowired
    private MyConfig myConfig;
    @RequestMapping(value = "/getMyConfig")
    public String getMyConfig() {
        return myConfig.toString();
    }
    @RequestMapping(value = "/getUsername")
    public String getUsername() {
        return myConfig.getUsername();
    }
    @RequestMapping(value = "/getYear")
    public String getYear() {
        return myConfig.getMy().getYear();
    }
}
```

2.5.7 在启动类引用相关配置

本实例中启动类代码如下。由于 MyConfig.java 对象中使用了@ConfigurationProperties 注解获得了 JDBC 头部参数的数据（prefix = "jdbc"），启动类中的@EnableConfiguration- Properties 注解则允许微服务工程使用@ConfigurationProperties 注解。

```
package com.zfx;
import org.springframework.boot.SpringApplication;
import org.springframework.boot.autoconfigure.EnableAutoConfiguration;
import org.springframework.boot.autoconfigure.SpringBootApplication;
import org.springframework.boot.context.properties.EnableConfigurationProperties;
import org.springframework.cloud.client.discovery.EnableDiscoveryClient;
import org.springframework.web.bind.annotation.RequestMapping;
import org.springframework.web.bind.annotation.RestController;
import com.zfx.config.MyConfig;
@RestController
@EnableDiscoveryClient
@SpringBootApplication
@EnableConfigurationProperties({MyConfig.class})
public class ApplicationMain {
    @RequestMapping("/health")
    public String health() {
        return "200";
    }
    public static void main(String[] args) {
        SpringApplication.run(ApplicationMain.class, args);
    }
}
```

@EnableConfigurationProperties 注解：启动对@Configurationproperties 注解的支持。该注解需要输入已使用@ConfigurationProperties 注解的类反射。若用逗号分隔，则可输入多个。

@Configurationproperties 注解：外部化配置的注解。若需要绑定和接收外部的一些资源配置文件，则需要使用该注解。

2.5.8 当前项目结构

如图 2-15 所示，本实例新增了 ConfigController.java 类管理相关接口，使用 My.java 与 MyConfig.java 类进行对相关 Value 值的映射与返回。如果不使用 Java Bean 形式，使用@Value、@ConfigurationProperties 等注解就可获得相关 Value 值。

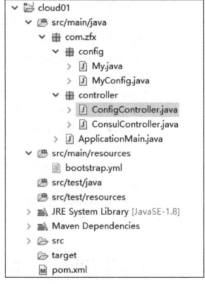

图 2-15

2.5.9 运行结果

1. 第一次获取相关参数

如图 2-16 所示，调用获得年份接口得到值 2019。

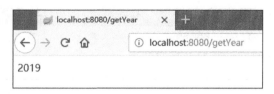

图 2-16

如图 2-17 所示，调用获得用户名接口得到值 myusername。

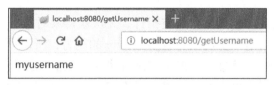

图 2-17

2．修改相关参数

如图 2-18 所示，在第一次调用接口获得参数后修改相关参数，并且使应用程序保持持续运行，可观察应用程序是否在未重启的情况下重新获得了新的参数。

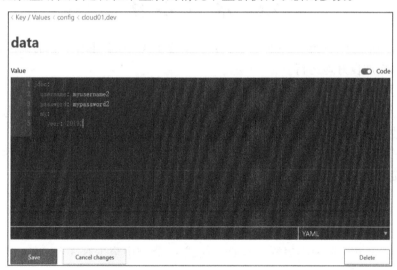

图 2-18

3．刷新 Controller 后第二次获取相关参数

如图 2-19 所示，重新调用获得年份接口得到新的值 20192。

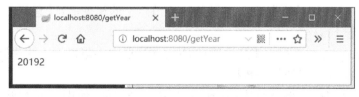

图 2-19

如图 2-20 所示，重新调用获得用户名接口得到新的值 myusername2。

图 2-20

2.5.10 实例易错点

本实例的易错点较多，包括抛出如空指针、无法获得某对象、某对象为空无法进行引用等相关错误。主要原因是本实例没有较多 Debug 的能力，所以可能导致项目报错过多，注意不要将一些重要的对应信息遗忘或出现配置错误，主要包括以下几点。

（1）YAML 资源配置文件中的每个参数是否与 Consul K/V 存储的 Key 地址相互对应。

（2）MyConfig 中的参数名称是否与 Consul K/V 存储的 Value 名称一一对应，数据格式是否一一对应。

（3）ApplicationMain 启动类是否开启了对引入配置注解的支持。

2.6 本章小结

2.3 节初步认识了 Consul 注册中心，并且搭建了 Consul 注册中心的集群，使 Consul 注册中心可以更加良好地运行。

2.4 节使用 Spring Cloud 整合了 Consul 进行注册并获得其他服务的注册地址。

2.5 节使用 Spring Cloud 整合了 Consul 的 Config 功能，可以通过 Java 代码获取 Consul 上的配置参数，方便多个微服务工程管理相同的配置信息。

在得到了其他微服务信息后，第 3 章将介绍使用 Feign 框架通过 Consul 注册中心调用其他微服务接口。

2.7 习题

1. 搭建 Consul 注册中心集群（非 Dev 模式），集群中至少有 3 台 Server 端（Linux 系统两台，Windows 系统 1 台）。

2. 创建 Spring Boot 工程，使其注册在 Consul 注册中心集群上（只注册一个地址），判断此时是否所有 Consul 注册中心节点都可以得到该 Spring Boot 工程的相关信息。

3. 创建 Spring Boot 工程，使其注册在 Consul 注册中心集群上（只注册一个地址），此时强行关闭该工程，记录所有 Consul 注册中心节点上的相关信息（包括 Linux 内部信息、命令行列表、日志）。

4. 创建 Spring Boot 工程，使其注册在 Consul 注册中心集群上，并获得该注册中心集群内其他 Spring Boot 工程的相关数据。

5. 在 Consul 注册中心集群的某 Consul 注册中心节点处，使用 K/V 存储保存相关数据，创建 Spring Boot 工程，获取保存在此 Consul 注册中心节点处的数据，并通过接口返回给前台。

第 3 章　分布式的通信

2.4 节中的实例已经获得了注册中心内其他服务的信息，此时可以直接通过 HTTP 使用 HttpClient、OKHttp、CXF、UrlOpenConnection 等方式对接口进行调用。但当服务数目增多后，如果使用上述方式调用接口，调用接口部分的代码将过于频繁与烦琐，而调用接口的前期步骤是几乎相同的代码，且不断复制 HttpClient 等相关的代码访问接口。

Spring Cloud 提供了 Feign 作为彼此通信的交流工具，只用一个注解就可以使用其他工程下提供的接口。

3.1　分布式通信

常见通信协议有对象访问协议 SOAP、超文本传输协议 HTTP、远程过程调用协议 RPC、传输控制协议 TCP、开放式系统互联协议 UDP 和全双工通信协议 WebSocket。其中，Spring Cloud 使用 HTTP。

为了使应用程序的编程更加简捷，一些基于协议上的应用级工具与框架应运而生，如 HttpClient、OKHttp、CXF、Netty 和 Mina，即不需要直接操纵协议就可以调用其他应用程序提供的服务，以此达到通信的目的。通常通信都需要相应的通信地址，分布式通信是指多个工程在同一个注册中心进行注册后，通过注册中心获得其他工程的通信地址，彼此进行远程服务的调用与访问。

3.1.1　Spring Cloud Feign

Spring Cloud Feign 是声明式 REST 风格调用工具，它使编写 Web 服务客户端更容易。微服务工程可以直接通过 Spring Cloud Feign 的相关注解对 HTTP 接口进行调用。被调用的 HTTP 接口不需要任何额外的注解修饰，如@RestController 或@Controller。

Spring Cloud Feign 集成了 Ribbon，提供了一个负载平衡的 HTTP 客户端。Spring Cloud Feign 支持可插拔的编码器和解码器，可以自定编码及解码格式。Spring Cloud Feign 底层默认基于 Java 自带的 UrlOpenConnection 进行调用，不需要额外依赖其他工程，并且允许更改 Spring Cloud Feign 的调用方式，如将 Java 自带的 UrlOpenConnection 改成 HttpClient 等第三方工具。

3.1.2　Swagger

在通信前，程序员 A 可能不知道程序员 B 写的微服务接口名、接口入参、接口返回参数等。不可能让每个程序员将每个微服务全看一遍，因此使用 Swagger 工具制作在线 API，并允许程序员 A 直接在 Swagger UI 上测试，节省了程序员 B 给程序员 A 制作 Word 文档或口述等的时间。

Swagger 可以生成在线文档，为用户和企业简化了 API 的开发，将接口可视化，并设置了相关的开源工具集。在 Swagger 的在线 API 生成过程中，只需要输出几行代码，就可以让用户直接使用 API，并通过在线 API 直接调用程序进行测试。Swagger 的工具集含有 API 治理、API 检测、API 的模拟与虚拟化、API 测试、在线 API 文档、离线 API 文档、API 设计等。

在微服务的分布式构建过程和多个微服务进行通信的过程中，如果利用 Word 文档告知彼此的入参、接口名、返回参数等相关内容，很容易出现版本不一致或出错的情况。虽然 Spring docs 可以生成离线文档，但是过程烦琐，而且每次更新都需要手动更新，文档重新传输。因此微服务的分布式构建通常利用 Swagger 作为解决方案。

3.2 【实例】微服务集成 Swagger

3.2.1 实例背景

本实例将使用微服务集成 Swagger，并在 Consul 中进行注册，为其他服务的调用做准备。本实例还将创建 cloud-user-8083 工程和 cloud-common-jar 工程。cloud-common 只作为依赖 Jar 包存在，并为其他工程提供相应的实体类，不整合 Boot 或 Cloud。限于篇幅本书不再介绍 cloud-common 工程。若在生产项目中，则应新建 cloud-parent 工程，规定依赖的版本号，将众多微服务的 Maven 依赖版本保持一致。

本书在工程命名的时候通常将项目的版本号写在工程名和启动类名里，只是为了测试时在 Eclipse 的 Console 控制台中显示更加直观。通常将工程名书写为 "项目名-微服务名-端口号"，启动类名书写为 "ApplicationMain" 或 "微服务名+AppMain"，配置文件中的 spring.application.name 分布式微服务名书写为 "项目名-微服务名" 等。注意，不可使用下画线 "_"，因为下画线在 Cloud 通信中可能会报错。

3.2.2 编写 Swagger 依赖

Swagger 需要引入 springfox-swagger2 和 springfox-swagger-ui 的依赖。pom.xml 文件所需增加的部分代码如下。

```
<dependency>
    <groupId>io.springfox</groupId>
    <artifactId>springfox-swagger2</artifactId>
    <version>2.7.0</version>
</dependency>
<dependency>
    <groupId>io.springfox</groupId>
    <artifactId>springfox-swagger-ui</artifactId>
```

```
    <version>2.7.0</version>
</dependency>
```

注意，Server 端不需要使用 Feign 进行依赖，声明式服务调用只需要在 Client 端进行依赖即可。

3.2.3　编写 Swagger 配置

创建 Swagger2.java 文件，Swagger2.java 文件用于 Swagger 的基本配置，代码如下。另外，Spring 的 YAML 资源配置文件与 2.3 节中的资源配置文件基本相同，不再赘述。

```
package com.zfx.config;
import org.springframework.context.annotation.Bean;
import org.springframework.context.annotation.Configuration;
import springfox.documentation.builders.ApiInfoBuilder;
import springfox.documentation.builders.PathSelectors;
import springfox.documentation.builders.RequestHandlerSelectors;
import springfox.documentation.service.ApiInfo;
import springfox.documentation.spi.DocumentationType;
import springfox.documentation.spring.web.plugins.Docket;
import springfox.documentation.swagger2.annotations.EnableSwagger2;
@Configuration
@EnableSwagger2
public class Swagger2 {
    public static final String SWAGGER_SCAN_BASE_PACKAGE = "com.zfx";
    public static final String VERSION = "1.0.0";
    @Bean
    public Docket createRestApi() {
        return new Docket(DocumentationType.SWAGGER_2)
                .apiInfo(apiInfo())
                .select()
                .apis(RequestHandlerSelectors.basePackage(SWAGGER_SCAN_BASE_PACKAGE))
                .paths(PathSelectors.any())
                .build();
    }
    private ApiInfo apiInfo() {
        return new ApiInfoBuilder()
                .title("分布式整合 Swagger 接口文档示例")
                .description("更多内容请关注：http://www.zfx.com")
                .version(VERSION)
                .termsOfServiceUrl("www.zfx.com")
                .contact("张方兴")
                .build();
    }
}
```

@Configuration 注解将 Swagger2 的相关配置注册到 Spring 容器中。

@EnableSwagger2 注解开启 Swagger2 的相关注解和功能,开启的相关注解包括 Swagger2 的@Api 和@ApiOperation 等。

在 Swagger2 的配置中,相当于将 Docket 注册到 Spring 容器中,Docket 是 Swagger2 的构造器,Swagger 是基于 Spring MVC 构建的,Docket 提供合理的默认值和方便的配置方法。Docket 对象可配置的参数及释义如表 3-1 所示。

表 3-1 Docket 对象可配置的参数及释义

参数	释义
apiInfo	apiInfo 中存储 Swagger 的元数据,可将 API 的元信息设置为包含在 json 资源列表响应中
select	启动用于 API 选择的生成器
apis	API 接口包扫描路径,通常扫描 Controller 接口或直接扫描根目录
paths	可以根据 URL 路径设置哪些请求加入文档,忽略哪些请求。输入的 PathSelectors.any()代表任何路径都满足此条件

apiInfo 可配置的参数及释义如表 3-2 所示。

表 3-2 apiInfo 可配置的参数及释义

参数	默认参数	释义
version	1.0	版本号
title	Api Documentation	API 标题
description	Api Documentation	API 注释
termsOfServiceUrl	urn:tos	服务 URL
contact	Contact Email	联系邮件地址
license	Apache 2.0	文档的 License 信息
licenseUrl	http://www.apache.org/licenses/LICENSE-2.0	文档的 License 地址

3.2.4　编写接口与接口处的 Swagger 配置

为方便观察运行效果,在 UserController.java 文件中编写部分用于测试的接口。每个接口都只返回一个固定参数并打印相关日志,代码如下。

```
package com.zfx.controller;
import org.springframework.web.bind.annotation.ModelAttribute;
import org.springframework.web.bind.annotation.PathVariable;
import org.springframework.web.bind.annotation.RequestBody;
import org.springframework.web.bind.annotation.RequestMapping;
import org.springframework.web.bind.annotation.RequestMethod;
import org.springframework.web.bind.annotation.RestController;
import com.zfx.entity.User;
import io.swagger.annotations.ApiImplicitParam;
import io.swagger.annotations.ApiImplicitParams;
import io.swagger.annotations.ApiOperation;
@RestController
public class UserController {
```

```java
@RequestMapping(value = "/getUser1", method = RequestMethod.POST)
@ApiOperation(value="获取用户方法1", notes="通过post请求获取用户",produces = "application/json",tags="UserController")
public User getUser1 (@RequestBody User user) {
    user.setAge(user.getAge());
    user.setName("已进入 cloud-user-8083 项目 getUser 当中");
    System.out.println("user.toString()"+user.toString());
    return user;
}
@RequestMapping(value = "/getUser2/{name}/{age}", method = RequestMethod.GET)
@ApiOperation(value="获取用户方法2", notes="通过get@PathVariable方式请求获取用户")
@ApiImplicitParams({
    @ApiImplicitParam(name = "name",value = "用户名",paramType = "path",dataType = "String"),
    @ApiImplicitParam(name = "age",value = "用户年龄",paramType = "path",dataType = "int")
})
public User getUser2 (@PathVariable String name,@PathVariable int age) {
    User user = new User();
    user.setAge(age);
    user.setName("已进入 cloud-user-8083 项目 getUser2 当中");
    System.out.println("user.toString()"+user.toString());
    return user;
}
@RequestMapping(value="/getUser3", method= RequestMethod.GET)
@ApiOperation(value="获取用户方法3", notes="获取用户方法3 notes",produces = "application/json")
public User getUser3(@ModelAttribute User user) {
    user.setAge(user.getAge());
    user.setName("已进入 cloud-user-8083 项目 getUser3 当中");
    System.out.println("user.toString()"+user.toString());
    return user;
}
}
```

UserController 类定义了3个接口，为了展示运行效果，每个接口都使用 Swagger 的描述性注解修饰，如下。

@ApiImplicitParams 注解：引入其他自定义的@ApiImplicitParam 资源。

@ApiImplicitParam 注解：强行自定义部分入参资源，也可使 Swagger 自动生成，如果自定义可多写描述信息。

@ApiOperation 注解：描述针对特定路径的操作或 HTTP 方法。HTTP 方法与路径的组合将创建一个唯一的操作。在注解不被定义的情况下 Swagger 也会运行，但为了给每个接口描述信息，会编写该注解。@ApiOperation 注解配置的参数及释义如表3-3所示。

表3-3　@ApiOperation 注解配置的参数及释义

配置参数	类型	释义
Value	String	提供此操作的简要说明，应不超过120个字符，以便在用户界面中获得适当的可见性

续表

参　数	类　型	释　义
notes	String	对操作的详细描述
tags	String[]	作为文档索引，可根据 tags 值进行分组
response	Class<?>	返回值类型，Swagger 会将 user.class 转换成 Swagger 自身的 Model
httpMethod	String	请求类型，包括"GET"、"HEAD"、"POST"、"PUT"、"DELETE"、"OPTIONS"和"PATCH"
produces	String	响应参数类型，默认为空，application/json 为 json 输出，application/xml 为 XML 输出
authorizations	Authorization[]	高级特性认证时配置
hidden	Boolean	是否隐藏该 API，默认为 False
responseHeaders	ResponseHeader[]	响应头内参数的数组
code	int	必须是 HTTP 状态码，默认为 200

3.2.5　当前项目结构

启动类与 2.3 节中的启动类相同。

图 3-1 的 cloud-user-8083 工程依赖于图 3-2 的 cloud-common-jar 工程，本实例将运行 cloud-user-8083 工程，运行后观察 Swagger 生成的在线文档。

图 3-1

图 3-2

3.2.6　运行效果

1. 运行查看 Swagger UI 页面

运行后输入 Swagger 默认的浏览地址：http://localhost:8083/swagger-ui.html，可查看 Swagger 生成的页面，如图 3-3 所示。

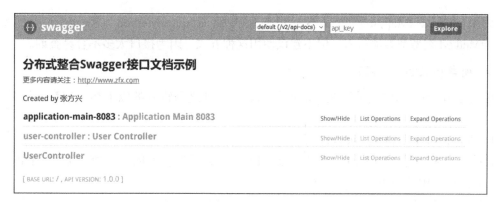

图 3-3

在 Swagger UI 页面中可观察目前所有 Swagger 扫描的接口，而且 Swagger2.java 类中 Swagger 配置的相关文档注释都作为标题等元素显示在 Swagger UI 中。

2．观察 UserController 中被 Swagger 扫描到的接口

由于 getUser1 接口的@ApiOperation 注解中设置了 tags 参数"UserController"，所以 Swagger 默认将该接口设置了自己的分组。从页面上可以看出，如果不设置 tags，Swagger 会以 Controller.class 进行默认分组，如图 3-4 所示。

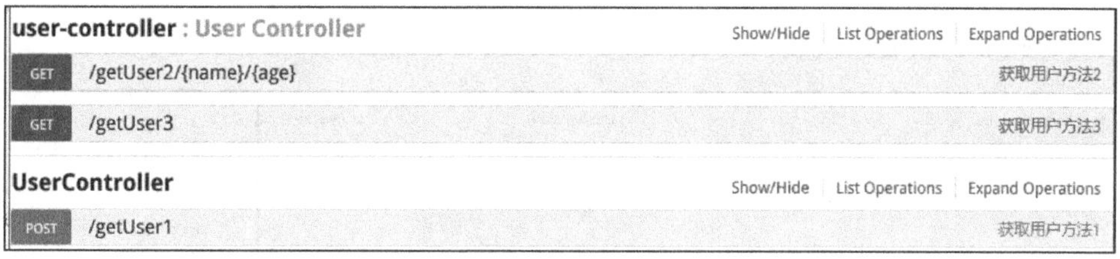

图 3-4

3．观察没有被@ApiOperation 注解修饰的接口

虽然没有在 ApplicationMain8083.class 启动类中编写@ApiOperation，但由于 Swagger 默认扫描该工程下的全部接口，所以 Consul 健康检查接口也被写入 Swagger UI 中，如图 3-5 所示。

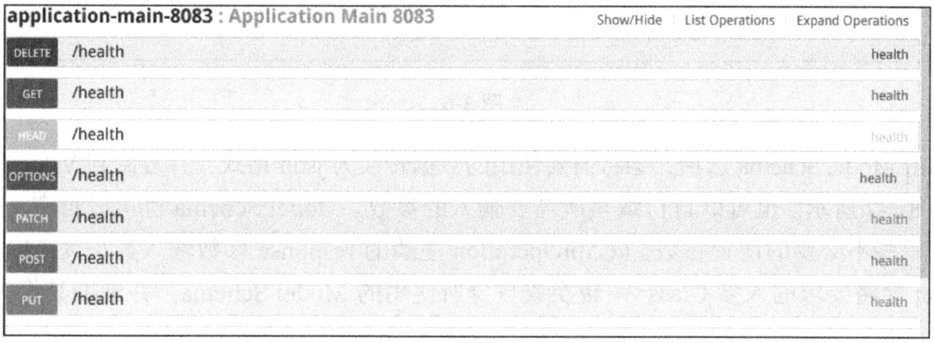

图 3-5

如果 @RequestMapping 注解不标识使用哪种 RequestMethod，Swagger 中的每个 RequestMethod 类型都会被写入，但不建议使用这种方式，因为接口太多不容易查询。

4．观察 getUser1 接口

单击 getUser1 选项显示该接口的详细列表，包括之前输入的以下相关信息，如图 3-6 所示。

Implementation Notes：操作详情。

Response Class：响应的 HTTP 状态码和返回参数类型，返回的参数是 User 对象。

Parameters 中的 Value：入参 json。

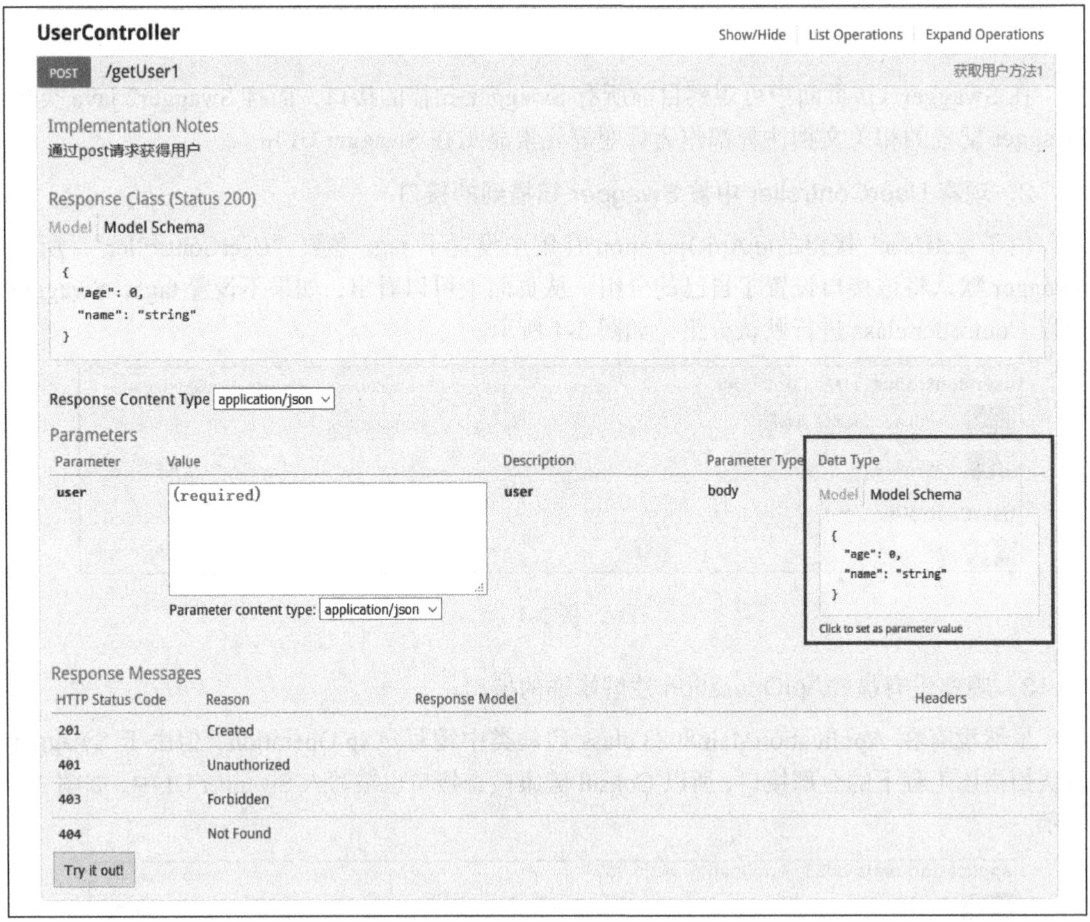

图 3-6

单击 Model Schema 选项，会将需要使用的数据转换为 json 格式，并复制到 Value 文本框中，如图 3-7 所示。也可以自行编写所需要输入的参数。Model Schema 也可以自定义格式，以减少一些不必要的内置参数。@ApiOperation 注解的 response 参数输入类型为 Class<?>，Swagger 会将要求的入参 Class<?>转换成自身所使用的 Model Schema，并辅助其作为入参使用。

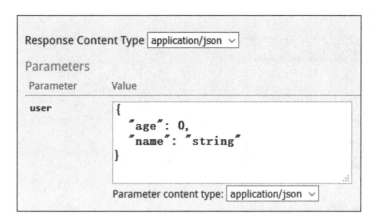

图 3-7

5．调用 getUser1 接口

更改 Parameter 入参中的 Value 后，单击 Try it out!按钮即可调用 getUser1 接口，如图 3-8 所示。

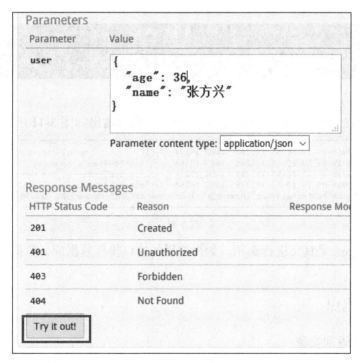

图 3-8

返回结果如图 3-9 所示。

Curl 文本框中的内容代表 Swagger 生成了利用 Linux Curl 工具调用 getUser1 接口的命令和调用该接口需要的入参，利用 Linux 控制台直接复制内容，调用结果如图 3-10 所示。

```
Curl

curl -X POST --header "Content-Type: application/json" --header "Accept: application/json" -d "{
  \"age\": 36,
  \"name\": \"张方兴\"
}" "http://localhost:8083/getUser1"

Request URL

http://localhost:8083/getUser1

Response Body

{
  "name": "已进入cloud-book-8083项目getUser当中",
  "age": 36
}

Response Code

200

Response Headers

{
  "content-type": "application/json;charset=UTF-8",
  "date": "Sun, 19 May 2019 09:17:23 GMT",
  "transfer-encoding": "chunked"
}
```

图 3-9

```
[root@localhost ~]# curl -X POST --header "Content-Type: application/json" --header "Accept: application/json" -d "{
> \"age\": 123,
> \"name\": \"123123123\"
> }" http://192.168.138.1:8083/getUser
{"name":"已进入cloud-book-8083项目getUser当中","age":123}[root@localhost ~]#
```

图 3-10

通过返回的响应体 name，调用已经到后台了，后台输出如图 3-11 所示。

```
2019-05-19 16:55:15.133  INFO 19164 --- [           main] com.zfx.A
2019-05-19 16:55:15.261  INFO 19164 --- [nio-8083-exec-2] o.a.c.c.C
2019-05-19 16:55:15.261  INFO 19164 --- [nio-8083-exec-2] o.s.web.s
2019-05-19 16:55:15.276  INFO 19164 --- [nio-8083-exec-2] o.s.web.s
user.toString()User [name=已进入cloud-book-8083项目getUser当中, age=36]
```

图 3-11

注意，Swagger 是真实进行调用，如果调用的接口操作数据库，数据库也会直接进行操作。

3.2.7　实例易错点

1．注解中的返回类型

@ApiOperation 注解的 produces 参数写成 application/json1，可视化页面如图 3-12 所示。

图 3-12

HTTP 无法识别此类型，即使传入参数也会出现相应错误，如图 3-13 所示。

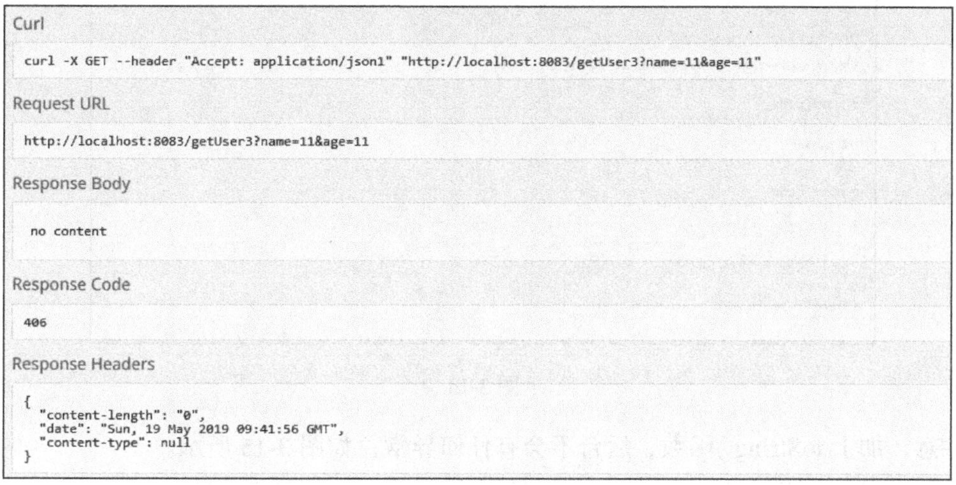

```
Curl
curl -X GET --header "Accept: application/json1" "http://localhost:8083/getUser3?name=11&age=11"

Request URL
http://localhost:8083/getUser3?name=11&age=11

Response Body
  no content

Response Code
406

Response Headers
{
  "content-length": "0",
  "date": "Sun, 19 May 2019 09:41:56 GMT",
  "content-type": null
}
```

图 3-13

HTTP 无法受理这些类型，除非自定义其编码格式。HTTP 的传输编码格式如表 3-4 所示。

表 3-4　HTTP 的传输编码格式

参　　数	释　　义
text/html	html 格式
text/plain	纯文本格式
text/xml	xml 格式
image/gif	gif 图片格式
image/jpeg	jpg 图片格式
image/png	png 图片格式
application/xml	xml 数据格式
application/json	json 数据格式
application/pdf	pdf 格式
application/msword	Word 文档格式
application/octet-stream	二进制流数据格式
multipart/form-data	文件上传

2．返回值不可加 toString()函数

因为使用@RestController 注解，其内部含有@ResponseBody 注解，所以可以自动将返回值转化成 json 类型。若返回时再加上 toString()，则 Swagger UI 无法正常调用。由于调用增加了 toString()的 getUser3 接口，所报错误如图 3-14 所示。

此处也会产生一个冲突问题，如果用 Feign 调用其他接口，其他接口若返回 toString() 后的参数，Feign 相关内容都不会报错，但 Swagger 会报错。因此为了程序统一，都不加 toString()函数，以避免出现此类错误。

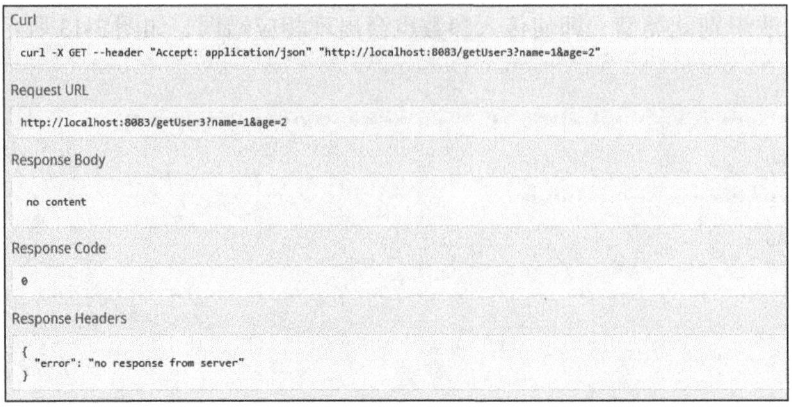

```
Curl
curl -X GET --header "Accept: application/json" "http://localhost:8083/getUser3?name=1&age=2"
Request URL
http://localhost:8083/getUser3?name=1&age=2
Response Body
  no content
Response Code
  0
Response Headers
{
  "error": "no response from server"
}
```

图 3-14

注意，加上 toString()函数，后台不会有任何异常，如图 3-15 所示。

```
2019-05-19 18:04:35.433  INFO 4108 --- [           main] o.s.b.w.embed
2019-05-19 18:04:35.441  INFO 4108 --- [           main] o.s.c.c.s.Con
2019-05-19 18:04:35.452  INFO 4108 --- [           main] com.zfx.Appli
2019-05-19 18:04:39.065  INFO 4108 --- [nio-8083-exec-2] o.a.c.c.C.[To
2019-05-19 18:04:39.065  INFO 4108 --- [nio-8083-exec-2] o.s.web.servl
2019-05-19 18:04:39.079  INFO 4108 --- [nio-8083-exec-2] o.s.web.servl
user.toString()User [name=已进入cloud-book-8083项目getUser3当中, age=2]
```

图 3-15

3．Swagger 的常见注解

表 3-5 所示为 Swagger 的常见注解及释义。

表 3-5 Swagger 的常见注解及释义

注　　解	释　　义
@Api	用于类，表示该类是 Swagger 资源
@ApiOperation	表示一个 HTTP 请求的操作
@ApiParam()	用于方法、参数、字段说明
@ApiModel()	用于类，表示对类进行说明，参数用实体类接收
@ApiModelProperty()	用于方法、字段，表示对 model 属性的说明
@ApiIgnore()	用于类、方法、方法参数，表示该方法或类被忽略
@ApiImplicitParam()	用于方法，表示单独的请求参数
@ApiImplicitParams()	用于方法，包含多个 @ApiImplicitParam

3.3 【实例】Feign 调用微服务接口

3.3.1 实例背景

目前已有 cloud-user-8083 工程和 cloud-common-jar 工程，本实例将创建 cloud-admin-

8084 工程，此工程同样引入 common-jar 工程，然后用 cloud-admin-8084 工程调用 cloud-user-8083 工程中的相关接口。

cloud-admin-8084 工程的 application 资源配置文件与 cloud-user-8083 工程中的资源配置文件基本相同，可参考 2.3.5 节，注意修改 spring.application.name 微服务名称即可。如果 spring.application.name 微服务名称相同导致重名，就会找不到相关服务，从而无法正常运行。

3.3.2 引入相关配置信息

虽然依赖 spring-cloud-starter-feign 和 feign-core 的结果基本相同，但依旧推荐依赖 spring-cloud-starter-feign，因为此处含有许多 netflix 其他相关依赖内容与组件，能够减轻依赖上的负担，并且防止出现的一些未知异常。

pom.xml 文件部分新增代码如下。

```xml
<dependency>
    <groupId>org.springframework.cloud</groupId>
    <artifactId>spring-cloud-starter-feign</artifactId>
    <version>1.4.0.RELEASE</version>
</dependency>
<dependency>
```

spring-cloud-starter-feign 的内部依赖代码如下。

```xml
<dependency>
    <groupId>io.github.openfeign</groupId>
    <artifactId>feign-core</artifactId>
</dependency>
<dependency>
    <groupId>io.github.openfeign</groupId>
    <artifactId>feign-slf4j</artifactId>
</dependency>
<dependency>
    <groupId>io.github.openfeign</groupId>
    <artifactId>feign-hystrix</artifactId>
</dependency>
```

3.3.3 编写 Feign 客户端

通过@FeignClient 注解，Feign 定义该 IndexService 对应 Consul 中注册的某个微服务，Value 参数输入微服务的名称，即提供服务工程的 spring.application.name，并在 IndexService.java 中编写 cloud-user 微服务的相关接口，然后在其他 Service 中直接注入 IndexService，即可获得 cloud-user 微服务的相关服务。

编写 Feign 客户端对应 cloud-user 微服务的 IndexService 代码如下。

```
package com.zfx.service;
import org.springframework.cloud.openfeign.FeignClient;
import org.springframework.web.bind.annotation.RequestBody;
import org.springframework.web.bind.annotation.RequestMapping;
import org.springframework.web.bind.annotation.RequestMethod;
import com.zfx.entity.User;
@FeignClient(value = "cloud-user")
public interface IndexService {
    @RequestMapping(value = "/getUser1", method= RequestMethod.POST)
    public User getUser1(User user);
}
```

@FeignClient 注解配置的参数及释义如表 3-6 所示。

表 3-6 @FeignClient 注解配置的参数及释义

配置参数	释义
Value	需调用的微服务名称，同时作为@FeignClient 实例的名称
Url	手动配置需要调用的微服务地址
decode404()	发生 404 后是否解码，而不抛出异常，默认为 false；若为 true，则解码，并不抛出异常
fallback	Feign 默认降级回退处理，一旦被调用接口发生异常，就回滚到某一个函数中
fallbackFactory	Feign 回滚地址的工厂，可以减少重复代码，增加重用性
configuration	@FeignClient 实例化的配置属性，包括 Encoder 编码器、Decoder 解码器、LogLevel 日志级别及 Contract 第三方注解翻译器

如果使用@GET/@POST 等第三方 JAXRS 注解，需要告知@FeignClient 正在使用第三方 JAXRS 注解，并指定第三方注解翻译器才可使用 JAXRS 注解。指定翻译器需在 Contract 中编写 new JAXRSContract()。

@RequestMapping 注解是 Feign 用来对应定义的 HTTP 接口地址，在 REST 风格下使用@GET、@POST 等相关注解也可。

Feign 提供了@RequestLine 注解用于对应 HTTP 接口，其作用等效于@RequestMapping，写法如下。

```
@RequestLine("GET /getUser2")
@Headers("Content-Type: application/json")
public String getUser2();
```

@Headers 注解是 Feign 用来增加报文头内部信息的注解。输入 String[]数组可用来输入多个 HTTP 报文头的信息，此处采用 json 编码格式。

@RequestLine 注解常用写法如下。

```
@RequestLine("POST /servers")
public void model();

@RequestLine("GET /servers/{serverId}?count={count}")
```

```
public void model(@Param("serverId") String serverId, @Param("count") int count);

RequestLine("GET")
public Response model(URI nextLink);

@RequestLine("GET /servers/{serverId}?count={count}")
public void model(@Param("serverId") String serverId, @Param("count") int count);

@RequestLine("GET")
public Response model(URI nextLink);
```

注意，在上述代码中 GET/POST 与后面的映射路径间有一个空格。

3.3.4 编写调用

编写 cloud-admin-8084 工程的 IndexController.java 文件代码如下，用于 cloud-admin-8084 工程调用 cloud-user-8083 工程的相关服务。

```
package com.zfx.controller;
import org.springframework.beans.factory.annotation.Autowired;
import org.springframework.web.bind.annotation.RequestMapping;
import org.springframework.web.bind.annotation.RestController;
import com.zfx.entity.User;
import com.zfx.service.IndexService;
@RestController
public class IndexController {
    @Autowired
    private IndexService indexService;
    @RequestMapping("/index")
    public String index() {
        User user = new User();
        user.setAge(10);
        user.setName("张方兴");
        return indexService.getUser1(user).toString();
    }
}
```

因为 Feign 对调用的接口进行了实例化，所以调用服务的相关代码，可直接将 indexService 通过 Spring 容器注入 Controller 中使用。若用实现的方式写上述代码，伪代码如下。

```
package com.zfx.test;
import feign.Headers;
import feign.RequestLine;
public interface FeignClient {
    @RequestLine("POST /getUser")
```

```
        @Headers("Content-Type: application/json")
        public String getUser();
    }
package com.zfx.test;
import feign.Feign;
public class FeignTest {
    public static void main(String[] args) {
        FeignClient feignClient = Feign.builder().target(FeignClient.class,"http://localhost:8083");
        String user = feignClient.getUser();
        System.out.println(user);
    }
}
```

注意，以上伪代码尽量不要在微服务分布式群中微服务互相调用时使用，否则日后维护会十分困难。

3.3.5 编写启动类

Feign 的客户端需要使用@EnableFeignClient 注解来启动 Feign 客户端的相关注解能力。ApplicationMain8084.java 启动类文件的代码如下。

```
package com.zfx;
import org.springframework.boot.SpringApplication;
import org.springframework.boot.autoconfigure.SpringBootApplication;
import org.springframework.cloud.client.discovery.EnableDiscoveryClient;
import org.springframework.cloud.openfeign.EnableFeignClients;
import org.springframework.web.bind.annotation.RequestMapping;
import org.springframework.web.bind.annotation.RestController;
@RestController
@EnableDiscoveryClient
@EnableFeignClients
@SpringBootApplication
public class ApplicationMain8084 {
    @RequestMapping("/health")
    public String health() {
        return "200";
    }
    public static void main(String[] args) {
        SpringApplication.run(ApplicationMain8084.class, args);
    }
}
```

3.3.6 当前项目结构

图 3-16 中的 cloud-admin-8084 工程，通过 Feign Client 请求图 3-17 中 cloud-user-8083 工程的@RestController 接口。

图 3-16   图 3-17

图 3-16 中的 cloud-admin-8084 工程与图 3-17 中的 cloud-user-8083 工程都依赖图 3-18 中的 cloud-common-jar 工程。

3.3.7 运行结果

运行 cloud-admin-8084 工程和 cloud-user- 8083 工程后，使用 cloud-admin-8084 的接口，可以看到 index 接口已经调用了 cloud-user-8083 的 getUser 接口。由此实现了通过一个注解让两个微服务进行通信，运行结果如图 3-19 所示。

图 3-18

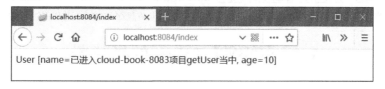

图 3-19

后台日志输出片段如下。

```
c.netflix.loadbalancer.BaseLoadBalancer: Client: cloud-user instantiated a LoadBalancer
```

netflix 在调用之前，依靠 BaseLoadBalancer 负载均衡器实例化了 Feign 服务的客户端。BaseLoadBalancer 相当于"服务池"，其内部维护了一个相对完整的服务列表。若用户需要调用注册中心的任何服务，则需要通过 BaseLoadBalancer 实例化服务。BaseLoadBalancer 服务池中也可以设置 Ping 心跳，以确定"服务列表"中的服务是否正常使用。

3.3.8 实例易错点

1. 没有找到响应的微服务

如果在编写@FeignClient(value = "cloud-user")注解时，value 内部对应的微服务名称填

写有误,在微服务启动的过程中不会报错,但是在调用 Feign 服务时,会报如下错误。

> com.netflix.client.ClientException: Load balancer does not have available server for client: cloud-user
> com.netflix.loadbalancer.LoadBalancerContext.getServerFromLoadBalancer(LoadBalancerContext.java:483) ~[ribbon-loadbalancer-2.2.5.jar:2.2.5]

上述错误表示没有找到 cloud-user 微服务,此时只要检查哪里写了 cloud-user 即可。若确定微服务名称正确,则检查 cloud-user 微服务是否已经启动。

2. 不要用下画线作为微服务的名称

在 3.3 节的工程中使用减号"-"构建微服务的名称,如 cloud-user。如果使用下画线"_"构建微服务名称,在服务端微服务运行、启动及注册到 Consul 注册中心上都不会出现问题,一旦 Feign 的调用方调用含有"_"的工程时,可能因为字符转义或其他因素,在调用方启动项目时出现如下错误。

> java.lang.IllegalStateException: Service id not legal hostname (cloud_user)
> at org.springframework.util.Assert.state(Assert.java:73) ~[spring-core-5.0.7.RELEASE.jar:5.0.7.RELEASE]
> Closing org.springframework.boot.web.servlet.context.AnnotationConfigServletWebServerApplicationContext

3. Swagger 与 Feign 版本冲突

低版本 Swagger 与 Feign 可能冲突,导致出现 Spring 容器无法正常实例化 Service 等相关错误,此时升级 Swagger 到 2.7 以上版本即可,其错误如下。

> org.springframework.beans.factory.UnsatisfiedDependencyException: Error creating bean with name 'indexController': Unsatisfied dependency expressed through field 'indexService'; nested exception is org.springframework.beans.factory.BeanCreationException: Error creating bean with name 'com.zfx.service.IndexService': FactoryBean threw exception on object creation; nested exception is java.lang.NullPointerException

3.4 【实例】Feign 的拦截器

3.4.1 实例背景

Feign 的拦截器是在 Client 端进行编写的,Client 端在调用 Server 端前,会在自身内部进行 Feign 拦截。通常拦截是为了在 Request Header 报文头内部增加 token 之类的信息,使 Feign 在调用前统一处理请求/请求头/请求体。

本实例继续修改 3.3 节中的 cloud-admin-8084 工程,通过实现 Feign 的拦截器保证每次 Feign 的调用都会被拦截函数进行拦截。

3.4.2 在 cloud-admin-8084 工程中增加拦截器

通过实现 feign.RequestInterceptor 接口,来实现 Feign 的拦截器。使用@Configuration 注解将 Feign 的拦截器注册到 Spring 容器上。在编写 Feign 拦截器 Interceptor 拦截器实现类的过程中,需要重写并实现 apply 函数,编写完该函数后,每次 Feign 请求都会经过 AdminInterceptor 拦截器。

AdminInterceptor.java 拦截器的实现代码如下。

```java
package com.zfx.interceptor;
import org.springframework.context.annotation.Configuration;
import feign.RequestInterceptor;
import feign.RequestTemplate;
@Configuration
public class AdminInterceptor implements RequestInterceptor{
    @Override
    public void apply(RequestTemplate template) {
        template.header("Content-Type", "application/json");
        template.header("token", "123456");
        System.out.println("已进入 AdminInterceptor 拦截器之中");
    }
}
```

RequestTemplate 是请求模板，即非线程安全的模板模型，可任意修改，所以通过复杂构造函数进行实现，请求对象都通过请求模板进行实例化。请求模板是 Feign 拦截器最重要的部分，RequestTemplate 的使用方法非常简单，直接使用其中的函数即可。

3.4.3 当前项目结构

如图 3-20 所示，增加 Feign 的拦截器只需要增加 AdminInterceptor.java 类即可，不需要增加其他内容。

3.4.4 运行结果

在调用请求时，先进入拦截器，然后通过 LoadBalancer 实例化 Feign 的负载均衡器，代码如下。

```
c.netflix.loadbalancer.BaseLoadBalancer : Client: cloud-user instantiated a LoadBalancer
```

客户端在实例化请求前，先进入拦截器，然后请求其他微服务，其后台日志如图 3-21 所示。

图 3-20

图 3-21

3.4.5 实例易错点

1. HTTP 编码格式错误

如果发生 HTTP 编码格式错误，控制台会输出错误信息如下。

com.netflix.client.ClientException: Load balancer does not have available server for client: cloud-user

此时需要在拦截器中添加 header 报文头，更改 HTTP 编码格式为

template.header("Content-Type", "application/json");

由于此处没有编辑器校验，所以容易写错，若写错，Feign 进入拦截器更改 HTTP 编码格式传输到 Server 端的时候，发生 Client 端的调用错误。注意，Feign 不接受 get 请求进行实体类 POJO 作为参数传递，解决方案如下。

（1）将 POJO 转换成 Map 进行传递。

（2）将 POJO 拆分成多个参数再进行传递。

（3）入参以@ModelAttribute 注解修饰，被@ModelAttribute 注解修饰后，POJO 会作为字符串拼接在 URL 上。但如果将 POJO 拆分成多个参数，URL 过长可能会影响性能。

若在通常情况下 Server 服务提供方和 Feign Client 调用方的入参、出参等函数完全相同，则不会出现较多问题，代码如下。

@RequestMapping(value = "/getUser1", method= RequestMethod.***POST***) **public** User getUser1(User user);	Feign Client 调用方

@RequestMapping(value = "/getUser1", method = RequestMethod.POST)
public User getUser1 (User user) **throws** Exception {
user.setAge(user.getAge());
<u>user</u>.setName("已进入 cloud-book-8083 项目 getUser 当中");
System.***out***.println("user.toString()"+user.toString());
return user;
}
	Server 服务提供方

2. Feign 拦截器需注册到 Spring 容器中

如果没有将 Feign 拦截器作为 Bean 注册到 Spring 容器中，相当于没有自定义拦截器，不会发生其他任何异常。若没有注册到 Spring 容器中，可用编程的方式实例化 FeignClient，伪代码如下。

FeignClient feignClient = Feign.builder()
    .requestInterceptor(new AdminInterceptor())
    .target(FeignClient.class,"http://localhost:8083");

通过代码构造了 Feign 的 Client 端，构造时已经将拦截器作为构造参数使用，因此不需要将 Feign 拦截器注册到 Spring 容器中。除上述方式外，将拦截器配置到 application.yml 资源配置文件中，也不需要将拦截器作为 Bean 注册到 Spring 容器中，配置代码如下。

```yaml
feign:
  client:
    config:
      #FeignClient 的名字
      cloud-user:
        #配置拦截器节点
        requestInterceptors:
          - com.zfx.interceptor.AdminInterceptor
          - xxx.xxx.xxxxxxxxxx.xxxxxxxxxxxx
```

此时需注意以下 4 点。

（1）feign.client.config.cloud-user.requestInterceptors 中对应的数据相当于一个数组，可以多个叠加。

（2）"–"减号后有一个空格，"–"减号前比 requestInterceptors 节点前多两个空格。

（3）当前 FeignClient 的名字为 cloud-user，上述配置是添加在 cloud-admin-8084 工程中的，但由于实例化的是 cloud-user 接口，所以 FeignClient 的名字为 cloud-user。若拦截器较复杂，则推荐用资源配置的写法。如果要设置全局配置，可以将 cloud-user 改为 default，即 feign.client.config.default 下为 Feign 拦截器的默认配置。

（4）不要将 Feign 拦截器注册到 Spring 容器中的同时，又在资源配置文件中进行配置，否则该场景下 Feign 会两次进入拦截器，输出日志如图 3-22 所示。

```
ApplicationMain8084 [Java Application] G:\tools\java1.8\JDK\bin\javaw.exe (2019年5月20日
已进入AdminInterceptor拦截器之中
已进入AdminInterceptor拦截器之中
2019-05-20 23:06:21.065  INFO 5860 --- [nio-8084-exec-1] s.
2019-05-20 23:06:21.109  INFO 5860 --- [nio-8084-exec-1] f.
2019-05-20 23:06:21.292  INFO 5860 --- [nio-8084-exec-1] c.
2019-05-20 23:06:21.316  INFO 5860 --- [nio-8084-exec-1] c.
```

图 3-22

3.5 Feign 的配置

Feign 在 Spring Boot 微服务的 application 资源配置文件中支持传输数据压缩配置、日志配置和超时配置。传输数据压缩配置可减少 Feign 的 Client 端调用接口所需的字节数量，加快传输速度。日志配置方便管理 Feign 自身调用的日志信息，在实际生产中可以减少日志存储量，并方便日志的归档、总结和管理。超时配置可定义在 Feign 的 Client 端调用接口时，Server 端接口多久不响应则 Feign 进行降级回退处理。降级回退可参见 3.6 节。

3.5.1 传输数据压缩配置

Feign 支持对响应进行 GZIP 压缩，用于提高 Server 端和 Client 端的通信效率，在相应的资源配置文件中进行配置即可，不需要进行额外操作，配置代码如下。

```yaml
feign:
  compression:
    request:
      # 开启请求压缩
      enabled: true
      # 配置压缩的 MIME TYPE
      mime-types: text/xml,application/xml,application/json
      # 配置压缩数据大小的下限
      min-request-size: 2048
    response:
      # 开启响应压缩
      enabled: true
```

3.5.2 日志配置

Feign 中的每个 @FeignClient 注解都会生成一个 FeignClient 实例，每个 FeignClient 实例都会对应一个 Feign.Logger 日志输出实例，每个 Feign.Logger 日志输出实例都会根据资源配置文件中规定的日志输出规范，进行日志输出，配置代码如下。

```yaml
logging:
  level:
    com.zfx.service.IndexService: none
```

none 下的日志输入如图 3-23 所示。

图 3-23

none 级别不会输出任何 Feign 相关的日志，图 3-23 中的日志是 System.out.print 输出的。Feign 通过不同的日志级别输出不同规范的日志。如表 3-7 所示，可以设置任意 FeignClient 日志级别。

表 3-7 FeignClient 的日志级别及释义

日志级别	释 义
none	不记录任何日志信息
basic	记录请求函数、URL、响应状态码、执行时间
headers	记录请求和响应的头部信息和 basic 级别信息
full	记录全部信息

Feign 默认使用 full 级别记录日志。

3.5.3 超时配置

Feign 可以通过内置的 Ribbon 负载均衡器设置超时时间，配置代码如下。

```
ribbon:
    ReadTimeout: 100 #请求处理的超时时间
    ConnectTime: 100 #请求连接的超时时间
```

以上时间单位均为毫秒，为了显示相应效果，此处设置 100 毫秒。如果 Feign 调用端设置了如上配置，Feign 的服务端可以增加相应的线程停止代码，方便测试超时后的效果，部分代码如下。

```
@RequestMapping(value = "/getUser1", method = RequestMethod.POST)
@ApiOperation(value="获取用户方法 1", notes="通过 post 请求获得用户",produces = "application/json",tags=
"UserController")
public User getUser1 (@RequestBody User user) throws Exception {
    user.setAge(user.getAge());
    user.setName("已进入 cloud-book-8083 项目 getUser 当中");
    System.out.println("user.toString()"+user.toString());
    Thread.sleep(300000);
    return user;
}
```

若每次服务调用该接口停留时间过长，Feign 则会直接进行降级回退处理，运行结果如图 3-24 所示。

图 3-24

虽然服务端依旧进行处理，但因调用端认为此响应时间超长而直接进行回退，并不因为接口错误进行回退。服务端日志如图 3-25 所示。

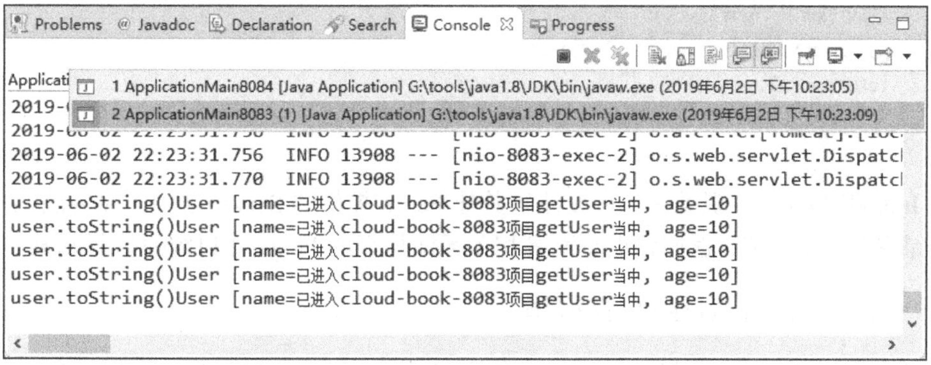

图 3-25

3.6 【实例】Feign 的降级回退处理——Feign 的 Fallback 类

3.6.1 实例背景

Feign 难免会遇到某个接口由于网络或项目宕机，从而无法正常调用其接口的情况。此时为了保证 Feign 的 Client 端不会出现异常，Feign 内置了 Hystrix 使用 Fallback 类作为容错解决方案。如果发生无法调用 Feign 提供端的情况，可以让其接口使用自定义的 Fallback 接口进行降级回退处理，保证应用程序的正常返回。

本实例继续修改 3.4 节的 cloud-admin-8084 工程，使用 Feign 的降级回退处理函数对异常进行容错与回退处理。

3.6.2 在资源配置文件中开启 Feign 内置的 Hystrix 权限

Feign 需要开启内置的 Hystrix 权限才可进行降级回退处理，配置代码如下。

```
feign:
  hystrix:
    enabled: true
```

3.6.3 编写 Fallback 降级类

IndexFallBack 降级类实现自 IndexService 接口，该接口已编写在 3.3 节中，一旦 getUser1 接口无法调用，都会使用 IndexFallBack 中的 getUser1 实现。IndexFallBack.java 降级实现类代码如下。

```java
package com.zfx.service.fallback;
import org.springframework.stereotype.Component;
import com.zfx.entity.User;
import com.zfx.service.IndexService;
@Component
public class IndexFallBack implements IndexService{
    @Override
    public User getUser1(User user) {
        User user2 = new User();
        user2.setName("fall back2 中的 user");
        return user2;
    }
}
```

在 IndexFallback 实现类中也可以注入其他 Feign 提供的 Server 端中，并再次调用。将重新登录账号密码、统一返回错误页面等相关无法执行的 Feign 接口统一执行到某一位置。

3.6.4 Service 整合 Fallback 降级类

通过@FeignClient 注解增加 Fallback 参数，并指向 IndexFallBack.class 降级实现类的映

射，整合 Fallback 降级类代码如下。

```
package com.zfx.service;
import org.springframework.cloud.openfeign.FeignClient;
import org.springframework.web.bind.annotation.RequestMapping;
import org.springframework.web.bind.annotation.RequestMethod;
import com.zfx.entity.User;
import com.zfx.service.fallback.IndexFallBack;
@FeignClient(value = "cloud-user",fallback=IndexFallBack.class)
public interface IndexService {
    @RequestMapping(value = "/getUser1", method= RequestMethod.POST)
    public User getUser1(User user);
}
```

3.6.5 当前项目结构

如图 3-26 所示，Feign 的降级回退处理只需要增加 IndexFallBack.java 即可，建议编写代码时，每个 FeignService 对应一个 FeignFallBack。

图 3-26

3.6.6 运行结果

只启动 cloud-admin-8084 工程而不启动 cloud-user-8083 工程，若调用 index 接口，在前文中则报 3.3.8 节中的错误。由于编写了 Fallback 降级类，所以 Feign 会调用 Fallback 中的类，并进行相关处理，再返回给前台，如图 3-27 所示。

另外，3.4 节中的拦截器类，由于采用先进入拦截器后进行请求的执行步骤，所以拦截器部分不会有任何异常，信息如下。

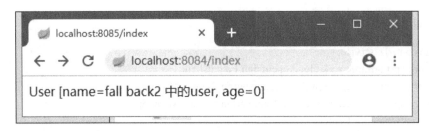

图 3-27

```
2019-06-02 17:52:13.551    INFO 9072 --- [ix-cloud-user-1] c.n.l.DynamicServerListLoadBalancer      :
DynamicServerListLoadBalancer for client cloud-user initialized: DynamicServerListLoadBalancer:{NFLoad-
Balancer:name=cloud-user,current list of Servers=[],Load balancer stats=Zone stats: {},Server stats: []ServerList:
ConsulServerList{serviceId='cloud-user', tag=null}
已进入 AdminInterceptor 拦截器之中
```

3.7 【实例】Feign 的降级回退处理——Feign 的 Fallback 工厂

3.7.1 实例背景

在 3.6 节中使用了 Fallback 类，处理 Feign 某一个 Service 端没有正常返回时的降级处理。Feign 也提供了 FallbackFactory<?>降级回退工厂，作为容错解决方案。

本实例继续修改 3.4 节中的 cloud-admin-8084 工程，使用 Feign 的降级回退工厂对异常进行容错与回退处理。在资源配置文件中开启 Feign 内置的 Hystrix 接口等部分代码参见 3.6 节。

3.7.2 编写 Fallback 降级工厂

编写 Fallback 降级工厂实现类 IndexFallBack.java 代码如下，Fallback 降级工厂类需要实现 FallbackFactory 工程接口，并重写 create 函数，如果 IndexFallBack.java 整合该降级工厂后，所有 IndexService 接口发生错误，就会使用 create 接口返回。

```java
package com.zfx.service.fallback;
import org.springframework.stereotype.Component;
import com.zfx.entity.User;
import com.zfx.service.IndexService;
import feign.hystrix.FallbackFactory;
@Component
public class IndexFallBack implements FallbackFactory<IndexService>{
    @Override
    public IndexService create(Throwable throwable) {
        return new IndexService() {
            @Override
            public User getUser1(User user) {
                User user2 = new User();
```

```
                user2.setName("fall back 中的 user");
                return user2;
            }
        };
    }
}
```

3.7.3 整合 Fallback 降级工厂

当前项目结构和运行结果与 3.6 节完全相同，IndexService.java 整合 Fallback 降级工厂代码如下。

```
package com.zfx.service;
import org.springframework.cloud.openfeign.FeignClient;
import org.springframework.web.bind.annotation.RequestMapping;
import org.springframework.web.bind.annotation.RequestMethod;
import com.zfx.entity.User;
import com.zfx.service.fallback.IndexFallBack;
@FeignClient(value = "cloud-user" , fallbackFactory = IndexFallBack.class)
public interface IndexService {
    @RequestMapping(value = "/getUser1", method= RequestMethod.POST)
    public User getUser1(User user);
}
```

3.7.4 实例易错点

1．未找到降级函数

Feign 降级代码十分简单。注意，IndexFallBack 类中需使用@Component 注解，将 IndexFallBack 类注入 Spring 容器中；在资源配置文件中需开启 Feign 内置 Hystrix 权限，否则会找不到 Fallback 类从而发生错误，报错信息如下。

com.netflix.client.ClientException: Load balancer does not have available server for client: cloud-user

此信息与 3.3.8 节中的错误完全相同。没有书写资源配置文件的开启信息或缺少注册到 Spring 容器中的注解，都会报上述错误。

2．降级函数初始化错误

如果 Feign Client A 引入实现 Feign Client B 的降级回退函数，会报初始化错误如下。

org.springframework.beans.factory.UnsatisfiedDependencyException: Error creating bean with name 'indexController5': Unsatisfied dependency expressed through field 'indexService2'; nested exception is org.springframework.beans.factory.BeanCreationException: Error creating bean with name 'com.zfx.service.IndexService2': FactoryBean threw exception on object creation; nested exception is java.lang.IllegalStateException: Incompatible fallback instance. Fallback/fallbackFactory of type class com.zfx.service.fallback.IndexFallBack is not assignable to interface com.zfx.service.IndexService2 for feign client cloud-user

解决方法是使 Feign Client 调用实现自身接口的 Feign 降级回退即可。

3.8 本章小结

基于微服务的分布式架构，本章使用 Feign 达到多个微服务互相通信的目的。当调用 Server 端出现异常时，通过 Feign 降级回退函数返回。

微服务集成 Swagger 减少了多个程序员之间的沟通成本，在某程序员提供 Swagger UI 后，其他程序员可以直接了解接口地址、名称、入参、返回值等信息。在得到其他接口的信息后，通过 Feign Client 端可以调用 Server 端微服务提供的接口。Feign 的拦截器在 Feign Client 端调用其他 Server 接口时，对本次请求统一处理。

如果 Feign Client 端调用的 Server 端发生了报错现象，Feign 的 Fallback 类和 Feign 的 Fallback 工厂分别用降级回退类与降级回退工厂，直接将本次请求通过降级函数回退给前台，不会造成线程的阻塞。

如果 Feign Client 端调用的 Server 端发生超时现象，可通过配置 Feign 内置的 Ribbon 负载均衡器进行解决。一旦 Server 端发生超时现象，Feign 都会直接将本次请求通过降级函数回退给前台，不会造成线程的阻塞。

3.9 习 题

1. 创建 Spring Boot 工程，使其整合 Swagger 框架，并返回相关数据。

2. 创建 Spring Boot 工程 1 与 Spring Boot 工程 2，使其注册在 Consul 注册中心集群上，使用 Feign 让 Spring Boot 工程 1 与 Spring Boot 工程 2 进行通信。在 Spring Boot 工程 1 与 Spring Boot 工程 2 通信前，拦截该通信的相关内容，对其进行更改后，再允许通信。

（1）在 Spring Boot 工程 1 与 Spring Boot 工程 2 通信的同时，应如何减少日志代码的输出数量。

（2）若此时要求 Spring Boot 工程 2 在返回通信前进行强制关机，程序有何反应。

（3）若此时要求 Spring Boot 工程 2 在超长时间后才进行返回，程序有何反应。

（4）若此时要求 Spring Boot 工程 1 分别传输大小为 1MB、10MB、100MB、1GB 的数据，程序有何反应，在高并发多线程的情况下记录在某型号计算机配置下 Feign 的传输速度、Feign 的成功率、丢包等相关信息。

3. 创建 Spring Boot 工程 1 与 Spring Boot 工程 2，使其注册在 Consul 注册中心集群上，使用 Feign 让 Spring Boot 工程 1 与 Spring Boot 工程 2 进行通信。

（1）在 Spring Boot 工程 1 中整合 Feign 的 Fallback 类保证降级回退，在 Spring Boot 工程 1 与 Spring Boot 工程 2 通信前，拦截该通信的相关内容，对其进行更改后，再允许通信。强制 Spring Boot 工程 2 进行报错，观察程序状态。

（2）在 Spring Boot 工程 1 中整合 Feign 的 FallbackFactory 降级工厂保证降级回退，在 Spring Boot 工程 1 与 Spring Boot 工程 2 通信前，拦截该通信的相关内容，对其进行更改后，再允许通信。强制 Spring Boot 工程 2 进行报错，观察程序状态。

第 4 章 分布式的客户端负载均衡

第 3 章使用 Spring Cloud Feign 进行分布式之间的通信，不过一旦出现并发过高的情况，Feign 就无法正常返回，需使用 Fallback 降级回退，将回退内容返回调用方。

超时的熔断和降级回退可以解决无法正常返回的问题，但不能增加程序的并发负载能力，Spring Cloud 使用 Ribbon 工具作为客户端负载均衡解决方案，通过将某个微服务多次部署在不同的服务器上，使多台服务器共同承受并发的压力，达到分散压力的效果。将多个相同的微服务工程部署在不同服务器上，Ribbon 将获得注册中心上某个微服务的多个服务地址，通过 Ribbon 自身的计算，合理地将当前请求使用 Feign 转发到某个提供服务的地址。

3.6 和 3.7 节中使用的 Feign 通过配置 Feign 内置的 Ribbon 负载均衡器进行超时熔断。虽然 Feign 内置了 Ribbon，但是给予用户的功能较少，所以还需要整合 Ribbon 客户端负载均衡器，来完善分布式系统。

4.1 负 载 均 衡

在使用多台服务器分摊请求的压力时，由某一个中间件集中接收请求，该中间件将请求分发给不同的服务器，再集中返回，以此减少每台服务器在同一时刻所接收的请求数目，达到减轻服务器压力的目的。

4.1.1 传统服务器端负载均衡

最早的负载均衡技术是通过 DNS 实现的，在某个 DNS 的名字中配置多个地址，因而查询到同一个名字的请求将可能被转发到不同的地址中。

传统情况下，请求通过 Nginx、Keepalived、Zookeeper、LVS、Apache 等相关服务装置/服务器，通过预定义的算法和策略转发到不同的 Java Web 运行服务器中。负载均衡的目的是提高程序的并发能力，用多台实体的服务器提供服务。

4.1.2 Ribbon 客户端负载均衡

Ribbon 是客户端负载均衡框架，客户端在发送请求前会根据算法在客户端进行计算，然后合理地分发到不同的服务器上，而传统的负载均衡是将请求发送到某个中间件装置后，中间件将请求分配到某台服务器上。Ribbon 的特点如下。

（1）Ribbon 基于 HTTP 和 TCP 编写，Spring Cloud Ribbon 通过封装 Netflix Ribbon 实现。

（2）Ribbon 在请求前，先计算需要发送到哪个服务器上。

（3）Spring Cloud Ribbon 与 Eureka、Consul、Zookeeper 等注册中心不同，Ribbon 不需要独立运行，只需要将 Spring Cloud Ribbon 框架在工程中进行依赖，并直接使用即可。

4.2 【实例】Feign 整合 Ribbon 分发请求

4.2.1 实例背景

本实例继续编写 3.3 节的实例，由 Ribbon 进行计算并分发请求。

3.3 节已经编写完成 cloud-common-jar、cloud-user-8083、cloud-admin-8084 三个工程。cloud-admin-8084 工程调用 cloud-user-8083 提供的接口，这两个工程都依赖 cloud-common-jar 包。本实例将创建 cloud-admin-8085-2 工程与 cloud-book-8086 工程，cloud-admin-8085-2 工程代码与 cloud-admin-8084 工程代码基本一致，只进行微调。cloud-admin-8085-2 工程与 cloud-book-8086 工程都需要依赖 cloud-common-jar 包，以获取相关实体类信息，不需要重复编写实体类。cloud-book-8086 工程会调用 cloud-admin 提供的接口，该请求可能传输到 cloud-admin-8085-2 或 cloud-admin-8084 的任意一个工程中，而 cloud-admin-8085-2 和 cloud-admin-8084 两个工程则会请求 cloud-user-8083 工程，最后返回给 cloud-book-8086 工程。

在实际生产的过程中，不会创建 cloud-admin-8085-2 工程，此处只为了方便编程测试，以及查看日志输出。实际部署时，只要将相同的工程微调后直接部署在不同的服务器上即可。如图 4-1 所示为当前项目所需要的全部工程，注意区分调用者和被调用者之间的关系。

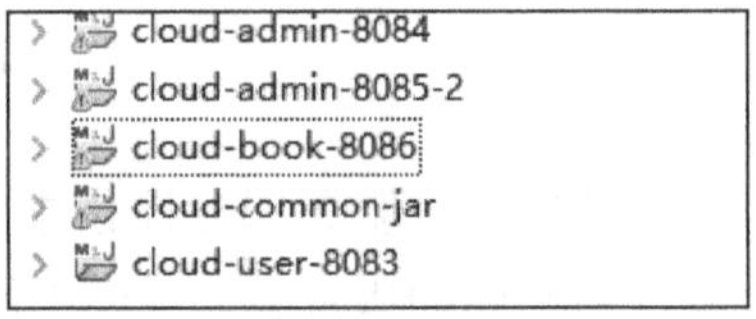

先改造 cloud-admin 工程的资源配置文件，cloud-admin-8084 工程的资源配置文件 application.yml 如下。

图 4-1

```yaml
spring:
  cloud:
    consul:
      host: localhost
      port: 8500
      discovery:
        healthCheckPath: /health
        healthCheckInterval: 15s
        hostname: localhost
  application:
    name: cloud-admin
feign:
  client:
    config:
      cloud-user-8083:
```

```yaml
      #连接超时
      connectionTime: 1000
      #读取超时
      readTimeout: 1000
      #配置拦截器节点
      requestInterceptors:
        - com.zfx.interceptor.AdminInterceptor
server:
  port: 8084
```

cloud-admin-8085 工程的资源配置文件 application.yml 如下。

```yaml
spring:
  cloud:
    consul:
      host: localhost
      port: 8500
      discovery:
        healthCheckPath: /health
        healthCheckInterval: 15s
        hostname: localhost
  application:
    name: cloud-admin
feign:
  client:
    config:
      cloud-user-8083:
        #连接超时
        connectionTime: 1000
        #读取超时
        readTimeout: 1000
        #配置拦截器节点
        requestInterceptors:
          - com.zfx.interceptor.AdminInterceptor
server:
  port: 8085
```

此处除更改了资源配置文件中的端口号外，还将两个微服务的名称统一改为 cloud-admin。使用 cloud-admin 名称注册在注册中心中后，注册中心发现某个服务由两个地址支持并提供，会将两个地址都存储到自身的空间中。

Ribbon 通过 Consul 或 Eureka 注册中心中记录的微服务名称，获取多个支持该微服务的地址，然后计算该微服务发送到哪个地址后，将请求发送到该微服务中。

4.2.2 编写 cloud-book-8086 启动类与配置类支持 Ribbon

使用@EnableFeignClients 注解开启对 Feign 相关注解的支持，并使用@RibbonClient 注解声明该微服务支持 Ribbon，启动类代码如下。

```
package com.zfx;
import org.springframework.boot.SpringApplication;
import org.springframework.boot.autoconfigure.SpringBootApplication;
import org.springframework.cloud.client.discovery.EnableDiscoveryClient;
import org.springframework.cloud.netflix.ribbon.RibbonClient;
import org.springframework.cloud.netflix.ribbon.RibbonClients;
import org.springframework.cloud.openfeign.EnableFeignClients;
import org.springframework.web.bind.annotation.RequestMapping;
import org.springframework.web.bind.annotation.RestController;
import com.zfx.config.ribbon.UserRibbonConfig;
@RestController
@EnableDiscoveryClient
@SpringBootApplication
@RibbonClients ({
        @RibbonClient(name = "cloud-admin", configuration = UserRibbonConfig.class)
})
public class ApplicationMain8086 {
    @RequestMapping("/health")
    public String health() {
        return "200";
    }
    public static void main(String[] args) {
        SpringApplication.run(ApplicationMain8086.class, args);
    }
}
```

@RibbonClient 注解：Ribbon 声明性配置类，该类中集成@Configuration 接口，注册在 Spring 容器中。在@RibbonClient 中需要输入微服务的名称，并且编写与其相对应的配置类信息。

将微服务的启动内容写到启动类中是为了方便未来维护，打开启动类即可知道该微服务启动和集成了哪些功能，并且统一在启动类中进行管理。也可以将@RibbonClients 放在单独的配置文件中，也可以写在统一的配置类里。应用程序通过 Ribbon 负载均衡器请求 cloud-admin 微服务的接口，该请求是根据 UserRibbonConfig.class 进行配置的。但是本实例先使用空白（默认）的配置 UserRibbonConfig.java，代码如下。

```
package com.zfx.config.ribbon;
@Configuration
public class UserRibbonConfig {}
```

4.2.3 Service 和 Controller

cloud-book-8086 微服务中的@Service 与@Controller 类与第 3 章中的类相同。

在 Feign 整合 Ribbon 的过程中,尽量不要用 Feign 编码形式或 org.springframework.web.client.RestTemplate 调用分布式微服务。

RestTemplate 错误的写法如下。

```
restTemplate.getForEntity("http://cloud-admin/getUser", User.class).getBody();
```

虽然以上部分代码确实可以调用微服务分布式，但是会给以后维护带来困难，在代码中需避免出现 String 形式的 URL 地址。如果分布式互相调用的代码较多，改动微服务的名称或返回类型，就需要更改其他几十个微服务中上百个调用地址，并且不确定是否全部的调用地址都已经修改。避免 Feign 通过编程的形式调用微服务接口，如使用 RestTemplate、HttpClient 等调用。通过编程形式调用 HTTP 接口，都只在调用微服务群外的 HTTP 接口时才使用。例如，工程需要调用 SMS 短信平台的 HTTP 接口，可以通过 Feign、RestTemplate、HttpClient 调用，而不在微服务群中调用。如果想用 HttpClient，可以更改 Feign 内置支持，将 Feign 使用的 UrlConnection 更改为 HttpClient。

Spring Cloud 只需要将 Ribbon 进行引用，并增加 Ribbon 的配置信息，Service 和 Controller 部分书写与 3.6 节一致即可。cloud-book-8086 工程的 IndexService.java 代码如下。

```
package com.zfx.service;
import org.springframework.cloud.openfeign.FeignClient;
import org.springframework.web.bind.annotation.RequestMapping;
import org.springframework.web.bind.annotation.RequestMethod;
@FeignClient(value = "cloud-admin")
public interface IndexService {
    @RequestMapping(value = "/index", method= RequestMethod.GET)
    public String index();
}
```

cloud-book-8086 工程的 IndexController.java 代码如下。

```
package com.zfx.controller;
import org.springframework.beans.factory.annotation.Autowired;
import org.springframework.web.bind.annotation.RequestMapping;
import org.springframework.web.bind.annotation.RestController;
import com.zfx.service.IndexService;
@RestController
public class IndexController {
    @Autowired
    private IndexService indexService;
    @RequestMapping(value="/index")
    public Object index() {
        String index = indexService.index();
        return index;
    }
}
```

上述为 Ribbon 的正确使用方式，Ribbon 只需在启动类引用即可，在 Controller 层和 Service 层不需要其他操作。

Feign 的编程式代码编写也应尽量减少，推荐使用注解和注入。微服务的分布式是为了简化开发压力并构建大型系统。上述 RestTemplate 形式若大量写在分布式中，开发压力更大。Rest 模板形式的写法了解即可。

4.2.4 当前项目结构

此时的微服务数量开始激增，图 4-2 和图 4-3 中的 cloud-admin 微服务将被图 4-4 中的 cloud-book-8085 微服务调用，cloud-admin 微服务将会调用图 4-5 中的 cloud-user-8083 微服务。其他微服务依旧依赖 cloud-common-jar 工程，如果之前没有拆分 Java Bean，将会出现 4 个相同的 user.java 类，整个应用程序将难以维护。

图 4-2

图 4-3

图 4-4 中的 cloud-book-8086 微服务将集成 Ribbon，从目录结构可以看出，即使集成了 Ribbon，cloud-book-8086 微服务也只增加了一个 Ribbon 的配置类，结构上并无其他变化。

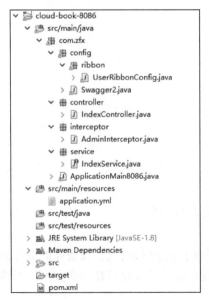

图 4-4

图 4-5

4.2.5 运行效果

本实例测试需要使用 cloud-book-8086 工程，利用 Ribbon 通过 8084 工程或 8085 工程的接口，调用 8083 工程的接口。

（1）启动 cloud-user-8083、cloud-admin-8084、cloud-admin-8085-2 三个工程，如图 4-6 所示。

图 4-6

（2）运行 cloud-book-8086 工程，执行 index 接口得到相应结果，如图 4-7 所示。

图 4-7

（3）在后台日志响应情况中 cloud-admin-8084 工程中的日志为空，如图 4-8 所示。

图 4-8

（4）cloud-admin-8085-2 工程中的日志有相应记录，如图 4-9 所示。

图 4-9

cloud-book-8086 工程通过 cloud-admin-8085-2 工程，调用了 cloud-user-8083 工程的接口。

（5）多次调用后后台日志响应情况如下。

多次调用后，会发现 cloud-admin-8085-2 工程和 cloud-admin-8084 工程都被调用了。cloud-admin-8084 工程日志响应情况如图 4-10 所示。

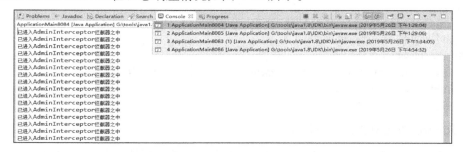

图 4-10

cloud-admin-8085-2 工程日志响应情况如图 4-11 所示。

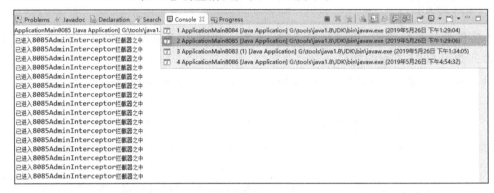

图 4-11

4.2.6 实例易错点

1．返回值类型错误

Feign 调用方异常，下述异常基本上是因为 Feign 调用方提供的服务返回值，与 Feign 提供方提供的服务返回值不同。

> org.springframework.web.client.RestClientException: Could not extract response: no suitable HttpMessageConverter found for response type [class com.zfx.entity.User] and content type [text/plain;charset=UTF-8]

如果 Feign 提供方代码如下，Feign 提供方提供 String 类型的返回值。

```
@Autowired
private IndexService indexService;
@RequestMapping(value="/index", method= RequestMethod.GET)
public String index() {
    User user = new User();
    user.setAge(10);
    user.setName("张方兴");
    return indexService.getUser1(user).toString();
}
```

如果 Feign 调用方使用其他返回值进行接收，会报此类异常，Feign 调用方代码如下也会报此异常。

```
@FeignClient(value = "cloud-admin")
public interface IndexService {
    @RequestMapping(value = "/index", method= RequestMethod.GET)
    public User index();
}
```

2．Ribbon 未找到匹配的微服务

Load balancer 是 Feign 内部集成 Ribbon 的负载均衡实现器，若找不到 cloud-admin 代

码，除将微服务名称写错外，在启动类集成 Ribbon 时所使用的名称不同也会报以上异常，代码如下。

> com.netflix.client.ClientException: Load balancer does not have available server for client: cloud-admin

Feign 调用方 Service 端代码如下。

```
package com.zfx.service;
import org.springframework.cloud.openfeign.FeignClient;
import org.springframework.web.bind.annotation.RequestMapping;
import org.springframework.web.bind.annotation.RequestMethod;
@FeignClient(value = "cloud-admin")
public interface IndexService {
    @RequestMapping(value = "/index", method= RequestMethod.GET)
    public String index();
}
```

Feign 调用方启动类集成 Ribbon 配置代码如下。

```
@RestController
@EnableFeignClients
@EnableDiscoveryClient
@SpringBootApplication
@RibbonClients ({
    @RibbonClient(name = "cloud-12323", configuration = UserRibbonConfig.class)
})
public class ApplicationMain8086 {...}
```

由于两处编写微服务的名称不同，Feign 匹配不上相关的微服务名称，所以会报此类异常。如果启动了@RibbonClients 注解，该工程中所有的 Feign 调用就需要在此处书写匹配的配置。

4.3　Ribbon 的负载均衡策略配置

Nginx 在默认情况下只进行轮询请求，即请求机器 1 一次之后，请求机器 2 一次，再循环。若机器 1 的性能优于机器 2，Nginx 可以采取权重加轮询的方式进行负载均衡，即机器 1 的权重高于机器 2，则优先将请求发往机器 1，并且机器 1 接受 n 次请求后，再去请求机器 2。这样可以保证在服务器性能不同的情况下，尽可能将性能压力进行平分。

4.2 节使用空的配置类，如果要更改 Ribbon 的负载均衡策略，只需要在配置类里增加相应 Spring Bean 即可，Ribbon 会在 Spring 容器中寻找相关 Bean。

UserRibbonConfig.java 配置类修改 Ribbon 策略代码如下。

```
package com.zfx.config.ribbon;
import org.springframework.context.annotation.Bean;
import org.springframework.context.annotation.Configuration;
import com.netflix.loadbalancer.IRule;
```

```
import com.netflix.loadbalancer.RoundRobinRule;
@Configuration
public class UserRibbonConfig {
    @Bean
    public IRule ribbonRule() {
        return new RoundRobinRule();
    }
}
```

在每次 UserFeign 调用 8084、8085 接口时，都会采取此策略调用 8084、8085 接口。Ribbon 策略及释义如表 4-1 所示。

表 4-1 Ribbon 策略及释义

Ribbon 策略	策略名称	释　　义
RoundRobinRule	轮询策略	依次轮询 Server
ResponseTimeWeightedRule	响应加权重策略	根据权重及 Server 响应时间选择 Server 调用
RandomRule	随机策略	随机选择 Server
RetryRule	重试策略	调用 Server 不成功后选择其他 Server 调用
BestAvailableRule	最佳可用性策略	选择并发数最低的 Server 调用
AvailabilityFilteringRule	可用性筛选策略	排除并发数最高的 Server 并调用其他 Server

Ribbon 默认采用 com.netflix.loadbalancer.ZoneAvoidanceRule 实现，该策略能够在多区域环境下选择最佳区域的实例访问。

4.4　本章小结

通过将多个微服务使用相同的微服务名称注册在同一个注册中心上，Feign Client 可以使用 Ribbon 根据算法调用其中任何一个微服务，通过多台服务器提高应用程序的并发承受能力。在分布式架构中，分布式通信是最重要的环节之一，注册中心保证分布式能够得到相关的通信地址，客户端负载均衡减小了分布式通信的并发压力。

4.5　习　　题

创建 Spring Boot 工程 1、Spring Boot 工程 2 与 Spring Boot 工程 2-2，使其注册在 Consul 注册中心集群上，使用 Feign 让 Spring Boot 工程 1 调用某接口，Spring Boot 工程 1 调用的接口由 Spring Boot 工程 2 和 Spring Boot 工程 2-2 同时提供，Spring Boot 工程 1 通过 Ribbon 随机调用 Spring Boot 工程 2 或 Spring Boot 工程 2-2 中的任意工程。

第 5 章 分布式的断路器

在第 2 章和第 3 章的实例中已经初步搭建了微服务的一个分布式程序，通过分布式的注册中心进行关联，并使用 Feign 工具进行通信，3.6 节与 3.7 节使用 Feign 内置的 Hystrix 进行降级回退处理。但由于 Feign 内置的 Hystrix 给予用户的功能较少，所以需要整合 Hystrix 断路器完善分布式系统的保护机制。

5.1 断 路 器

断路器是指分布式的保护机制，因为在分布式环境中，许多服务依赖项不可避免可能会失败。例如，在并发请求时，出现 TPS/QPS 过高、服务器宕机、服务器响应速度过慢、突发性网络波动异常等问题，因此需要保护机制进行处理。

5.1.1 为什么需要断路器

如果一套线路的电流过大，就会烧毁断路器中的保险丝，从而跳闸，一旦保险丝烧毁就保证过大的电流无法通过保险器，防止烧毁其他电器。Hystrix 断路器与物理中断路器的原理相似，即电流永远无法高于保险丝的承受能力。在分布式环境中，电流即并发，一旦调用并发太高，Hystrix 会开启断路器（跳闸），保证了微服务群中的其他微服务不会遭受损失。

如果集成 Hystrix，运维人员可根据 Hystrix 管理控制页面观察到目前分布式的运行情况，并且一旦出现任何突发性高并发，或 Feign 的 Server 突发性宕机、Feign 的 Server 返回时间过长导致阻塞大量消耗 CPU 与内存资源等情况，Hystrix 会自动进行熔断或降级处理，以保证程序正常运行。

5.1.2 Hystrix

Hystrix 是一个库，是分布式的一种保护机制，通过添加延迟容忍和容错逻辑来控制分布式服务之间的交互。Hystrix 通过隔离服务之间的访问点，停止跨服务的级联故障，以及提供后备选项做到容错、熔断、降级，提高系统整体的伸缩性与可用性。通过可视化 Hystrix 图表监控和 Hystrix 缓存分别控制分布式的维护难度和性能优化。

Hystrix 断路器除了防止用户因 Feign 的 Server 失败而导致错误，还监控函数的调用次数，若某函数的 Feign Client 被调用过多，引发 Feign Server 的报错率升高，Hystrix 会自动屏蔽过多的调用，以保证分布式系统的正常运行。

每个微服务运行都需要@SpringBootApplication 注解，而分布式微服务需要使用@SpringCloudApplication 注解，其中集成了启动注册中心注解@EnableDiscoveryClient、启动断路器注解@EnableCircuitBreaker 和启动微服务注解@SpringBootApplication。在分布式架构中，断路器、注册中心（包括通信）、微服务是微服务分布式的三大要素。Hystrix 保护机制与 Eureka 注册中心都由 Netflix 公司出品，且已于 2018 年 11 月停止了维护工作。但 Netflix 公司推荐 resilience4j 断路器作为新的保护机制方案。resilience4j 是受 Hystrix 启发而编写的断路器，通过管理远程调用的容错处理来帮助实现一个健壮的系统。resilience4j 提供了更好用的 API。虽然目前较少使用 resilience4j，但是断路器保护机制的技术更迭只是时间问题，未来 Hystrix 有可能被替换为 resilience4j。

对于新项目而言，更新断路器框架比较困难。国内除阿里巴巴等大型公司在使用自身编写的保护机制（如阿里巴巴的 Alibaba Sentinel 断路器）外，其他中小型公司都使用 Hystrix，暂时没有使用 resilience4j 断路器。而在注册中心方面已经有一些公司使用 Consul 了。resilience4j 部分操作与 Hystrix 基本相同。Hystrix 的可视化、缓存等功能，resilience4j 同样有。

Hystrix 使用@HystrixCommand(fallbackMethod="defaultMethod")作为降级回退。

resilience4j 使用@Retry(fallbackMethod = "defaultMethod")作为降级回退。

5.1.3 Hystrix 解决的问题

（1）防止任何单个依赖项耗尽所有容器（如 Tomcat）用户线程。

（2）若单个线程的执行时间过长，则直接让该线程快速失败，而不使多个线程持续排队。

（3）在可行的情况下提供降级回退处理，保证用户不受故障的影响。

（4）使用隔离技术（如断路器）限制任何依赖的影响。

（5）对分布式进行几乎实时的监控和报警，帮助用户优化程序。

（6）防止整个依赖客户端执行失败，而不仅仅在网络通信中。

5.1.4 Hystrix 如何解决问题

（1）将所有对 Feign 服务方（依赖项）的调用包装在 HystrixCommand 或 HystrixObservable Command 对象中，通过 Hystrix 的线程池让被包装的对象在单独的线程中执行。

（2）超时调用：HystrixCommand 对象使用独立运行的方式，独立运行所需时间若超过用户定义的阈值，独立线程则采用快速失败的方式停止运行。默认为 10 秒。

（3）为每个依赖项维护一个小线程池，若该依赖项已满，则发送给该依赖项的请求将立即被拒绝，而不是排队。默认情况下 10 秒 20 个请求以上且错误率达到 50%以上，Hystrix 就会自动开启断路器进行熔断，禁止其他微服务继续调用此函数。

（4）成功、失败（客户端引发的异常）、超时和线程拒绝。

（5）几乎实时地监控指标和配置变化。

（6）触发断路器以在一段时间内停止对某一特定服务的所有请求，若服务的错误百分比超过阈值，则手动或自动停止。

5.2 【实例】Hystrix 断路器的降级回退

5.2.1 实例背景

Hystrix 的降级回退与 3.6 节和 3.7 节的降级回退类似。在 Feign 的 Fallback 类中，需要重复实现多个 Service。Hystrix 的降级回退保证在某一个函数下，任何调用其他 Feign 服务端的接口出现错误，发生错误的函数将统一进行降级回退处理。

本实例将创建 cloud-book-8087-2 工程整合 Hystrix，测试时模拟函数执行异常，让程序进行降级回退处理。

5.2.2 编写相关 Pom 文件

由于本实例采用 Hystrix 断路器的熔断与回退，所以需要引入 Hystrix 的相关依赖，pom.xml 文件增加代码如下。

```xml
<dependency>
    <groupId>org.springframework.cloud</groupId>
    <artifactId>spring-cloud-starter-netflix-hystrix</artifactId>
</dependency>
```

5.2.3 编写 application 资源配置文件

application 资源配置文件与 4.2.1 节类似，Hystrix 断路器在使用过程中可以不增加任何配置信息，此时将采用 Hystrix 默认配置的方式执行。另外，Hystrix 断路器可以和 Feign 内置的 Hystrix 断路器同时使用。整合 Hystrix 断路器的 application.yml 资源配置文件代码如下。

```yaml
spring:
  cloud:
    Consul:
      host: localhost
      port: 8500
      discovery:
        healthCheckPath: /health
        healthCheckInterval: 15s
        hostname: localhost
  application:
    name: cloud-book
server:
  port: 8087
logging:
  level:
    com.zfx.service.IndexService: none
feign:
```

```yaml
hystrix:
    enabled: true
```

5.2.4　编写 Ribbon 配置类

Ribbon 配置类与 4.2 节的 Ribbon 配置相同，只要注册为 Spring Bean 即可。若有特殊需要可参考 4.3 节修改，Ribbon 策略配置类 UserRibbonConfig.java 代码如下。

```java
package com.zfx.config.ribbon;
import org.springframework.context.annotation.Bean;
import org.springframework.context.annotation.Configuration;
import com.netflix.loadbalancer.IRule;
import com.netflix.loadbalancer.RoundRobinRule;
@Configuration
public class UserRibbonConfig {
    @Bean
    public IRule ribbonRule() {
        return new RoundRobinRule();
    }
}
```

5.2.5　编写启动类

此时 @SpringCloudApplication 已经可以开始启用了。之前一直没有启用的原因是 @Spring-CloudApplication 注解中含有 @EnableDiscoveryClient、@EnableCircuitBreaker、@SpringBoot-Application 三个注解，在之前的实例中没有使用过断路器的相关功能，启用 @Spring-CloudApplication 注解会报错。而事实上每个微服务的分布式最基本的部分都需要微服务支持、注册中心（包括通信）支持和断路器支持。

若不是分布式类项目想使用 Hystrix 相关注解，则需要使用 @EnableHystrix 注解开启对断路器的支持。@EnableHystrix 注解中包含 @EnableCircuitBreaker 注解，用于启动 Hystrix 断路器。分布式启动类代码如下。

```java
package com.zfx;
import org.springframework.boot.SpringApplication;
import org.springframework.boot.autoconfigure.SpringBootApplication;
import org.springframework.cloud.client.discovery.EnableDiscoveryClient;
import org.springframework.cloud.netflix.hystrix.EnableHystrix;
import org.springframework.cloud.netflix.ribbon.RibbonClient;
import org.springframework.cloud.netflix.ribbon.RibbonClients;
import org.springframework.cloud.openfeign.EnableFeignClients;
import org.springframework.web.bind.annotation.RequestMapping;
import org.springframework.web.bind.annotation.RestController;
import com.zfx.config.ribbon.UserRibbonConfig;
//@EnableHystrix
//@EnableDiscoveryClient
//@SpringBootApplication
```

```java
@SpringCloudApplication
@RestController
@EnableFeignClients
@RibbonClients ({
        @RibbonClient(name = "cloud-admin", configuration = UserRibbonConfig.class)
})
public class ApplicationMain8086 {
  @RequestMapping("/health")
  public String health() {
        return "200";
  }
  public static void main(String[] args) {
        SpringApplication.run(ApplicationMain8086.class, args);
  }
}
```

5.2.6 编写 Service 类

Feign Client 的 Service 接口类与 3.3 节中的 Service 接口调用相同，Feign Client 的 Service 接口类代码如下。

```java
package com.zfx.service;
import org.springframework.cloud.openfeign.FeignClient;
import org.springframework.web.bind.annotation.RequestMapping;
import org.springframework.web.bind.annotation.RequestMethod;
import com.zfx.service.fallback.IndexFallBack;
@FeignClient(value = "cloud-admin")
public interface IndexService {
    @RequestMapping(value = "/index", method= RequestMethod.GET)
    public String index();
}
```

5.2.7 编写 Controller 类

为方便观察运行结果，可编写 IndexController.java 类，IndexController.java 代码如下。index 函数使用@HystrixCommand(fallbackMethod="defaultMethod")注解进行修饰后，在 index 接口中发生错误，都将降级回退到 defaultMethod()函数中。为了观察到降级回退结果，本实例故意使用数组越界错误。

```java
package com.zfx.controller;
import org.springframework.beans.factory.annotation.Autowired;
import org.springframework.web.bind.annotation.RequestMapping;
import org.springframework.web.bind.annotation.RestController;
import com.netflix.hystrix.contrib.javanica.annotation.HystrixCommand;
import com.zfx.service.IndexService;
@RestController
public class IndexController {
  @Autowired
```

```
private IndexService indexService;
@HystrixCommand(fallbackMethod="defaultMethod")
@RequestMapping(value="/index")
public Object index() {
    String index = indexService.index();
    int i = 1/0;
    return index;
}
private String defaultMethod(Throwable throwable) {
    System.out.println("已进入 private String defaultMethod()");
    System.out.println(throwable.getMessage());
    return "系统超时";
}
```

@HystrixCommand 注解：代表该函数将以 Hystrix 的线程池进行运行，需注意以下几点。

（1）如果 index 函数当前内部含有多个 Feign Client，只要其中有一个发生异常，系统就会整体回退到 defaultMethod 函数。

（2）在 defaultMethod 函数中，建议使用与 index 相同的入参、返回参数，以保证数据的完整性。

（3）在 index 函数中，如果 indexService.index() 执行成功后进行报错，那么 indexService.index();执行的 Feign Client 已经正常调用。如果该 Feign 的 Server 中含有增删改查，就会正常增删改查，不会进行回滚操作。

（4）此时如果外部调用 index 函数过多，或 index 函数内部中含有任何异常，或 indexService.index();返回时间过长，或 indexService.index();返回异常，那么程序都会回退使用 defaultMethod 函数。后续会对@HystrixCommand 注解进行各种优化与配置，Hystrix 常见的异常类型及释义如表 5-1 所示。

表 5-1　Hystrix 常见的异常类型及释义

异常类型	释义
THREAD_POOL_REJECTED	线程池拒绝
SHORT_CIRCUITED	强制启用断路器拒绝
TIMEOUT	超时拒绝
FAILURE	异常与失败拒绝
SEMAPHORE_REJECTED	信号量拒绝

5.2.8　当前项目结构

当前项目结构如图 5-1 所示，cloud-book-8087-2 微服务中的 Hystrix 断路器的降级回退被放置在@Controller 类中，所以与前文的项目结构相同。

图 5-1

5.2.9 运行结果

在 Hystrix 断路器降级回退的代码中，indexController 函数内部含有异常，程序运行结果如图 5-2 所示。

图 5-2

如果同时编写 @HystrixCommand 与 Feign 的 Fallback 类，@HystrixCommand 的实现类就更有可能返回给前台。此时并不一定能返回 @HystrixCommand 的降级实现，也有少量的测试表明 Feign 的 Fallback 降级类返回给前台，后台日志如图 5-3 所示。

图 5-3

无论 Server 端出现何种情况，Feign 永远先进入 Feign 的 Interceptor 拦截器，然后进入 @HystrixCommand 的实现方法，@HystrixCommand 降级回退函数和 Feign 的 Fallback 降级回退函数都会被调用。

5.2.10 实例易错点

1．HystrixCommand 回退函数的入参错误

5.2 节的 HystrixCommand 回退函数为了展示效果，入参写的是 Throwable。但在实际项目中，HystrixCommand 回退函数通常与 Service 或 Controller 函数中的入参相同。

HystrixCommand 回退函数与接口函数入参相同的伪代码如下。

```
package com.zfx.controller;
import org.springframework.beans.factory.annotation.Autowired;
import org.springframework.web.bind.annotation.RequestMapping;
import org.springframework.web.bind.annotation.RestController;
import com.netflix.hystrix.contrib.javanica.annotation.HystrixCommand;
import com.zfx.entity.User;
import com.zfx.service.IndexService;
@RestController
public class IndexController5 {
    @Autowired
    private IndexService indexService;
    @HystrixCommand()
    @RequestMapping(value="/index4")
    public Object myIndex(User user,String name) {
        User index = indexService.index();
        return index;
    }

    private String defaultMethod(User user,String name) {
        System.out.println("已进入 private String defaultMethod()");
        return "系统超时";
    }
}
```

如果 defaultMethod()函数与 index()函数的入参不同，可能会报如下错误。

```
@HystrixCommand(fallbackMethod="defaultMethod5")
@RequestMapping(value="/index5")
public Object myIndex(@ModelAttribute User user) {
    User index = indexService2.getUser1(user);
    return index;
}
private String defaultMethod5(User user,int i) {
    System.out.println("已进入 private String defaultMethod()");
    System.out.println(user);
```

```
        return "系统超时";
    }
```

com.netflix.hystrix.contrib.javanica.exception.FallbackDefinitionException: fallback method wasn't found: defaultMethod5([class com.zfx.entity.User])

在上述代码中,降级回退函数中多写了 int i,Hystrix 找不到入参为 User 的 defaultMethod5 函数,因此报如上错误,该错误即是找不到该降级回退函数。但如果入参正常,输入如下。

```
@HystrixCommand(fallbackMethod="defaultMethod5")
@RequestMapping(value="/index5")
public Object myIndex(@ModelAttribute User user) {
    User index = indexService2.getUser1(user);
    return index;
}

private String defaultMethod5(User user) {
    System.out.println("已进入 private String defaultMethod()");
    System.out.println(user);
    return "系统超时";
}
```

正常执行 HystrixCommand 熔断回退的日志如图 5-4 所示。注意,myIndex()函数中的入参 user 被传入 defaultMethod()函数的入参中。

```
已进入AdminInterceptor拦截器之中
2019-06-11 21:28:29.058  INFO 10496 --- [ix-cloud-use
2019-06-11 21:28:29.099  INFO 10496 --- [ix-cloud-use
2019-06-11 21:28:29.217  INFO 10496 --- [ix-cloud-use
2019-06-11 21:28:29.235  INFO 10496 --- [ix-cloud-use
2019-06-11 21:28:29.239  INFO 10496 --- [ix-cloud-use
已进入private String defaultMethod()
User [name=1, age=3333]
```

图 5-4

2. 没有找到 HystrixCommand 回退函数

如果只有 HystrixCommand 函数名出现错误,也会报相关错误,错误日志如下。

com.netflix.hystrix.contrib.javanica.exception.FallbackDefinitionException

错误代码如下。

```
@HystrixCommand(fallbackMethod="defaultMethod6")
@RequestMapping(value="/index5")
public Object myIndex(@ModelAttribute User user) {
    User index = indexService2.getUser1(user);
    return index;
}
```

```
private String defaultMethod5(User user) {
    System.out.println("已进入 private String defaultMethod()");
    System.out.println(user);
    return "系统超时";
}
```

5.3 Hystrix 线程池

5.2 节使用 Spring Cloud 整合了 Hystrix，做了一个基本的降级回退演示。每个函数通过@HystrixCommand 注解修饰后，在每次调用该函数时，Hystrix 都会创建相应的线程池，通过隔离的方式使用 Spring AOP 的@Around 环绕通知获得其中相关信息，包括 Fallback 回退位置与 Hystrix 相关配置等。@HystrixCommand 的 AOP 部分源代码在 com.netflix.hystrix.Contrib.javanica.aop.aspectj.HystrixCommandAspect 类中。有独立的线程在@Around 中运行，达到使用并监控整个函数线程的目的。

@HystrixCommand 的线程池由其注解内部的 groupKey 等多个参数整理和配置，groupKey 参数是 Hystrix 分配的线程组名称，默认情况下采用类名。除 groupKey 外，还有 commandKey 与 threadPoolKey。

commandKey 为 Hystrix 执行的命令名，默认情况下采用函数名，即每个函数都会被当成一个 HystrixCommand 执行。

threadPoolKey 默认情况下采用 groupKey 的名字，即采用该类的类名，也可自定义线程池的名字。threadPoolKey 用于从线程池缓存中获取线程池，并且初始化创建线程池。

commandKey 的 HystrixCommand 会在 threadPoolKey 中执行，若多个函数自定义配置 groupKey，并且 groupKey 相同，则多个函数都会在同一个线程池中运行。

若多个函数自定义配置的线程组名字 groupKey，与线程的名字 threadPoolKey 完全相同，则多个函数会在同一个线程池中执行但不推荐。groupKey 名与 threadPoolKey 名完全相同，因为这样不易定义线程数量、被调用次数和超时时间。例如，默认情况下每个函数 10 秒内被调用 20 次，并且报错率达到 50%以上，Hystrix 则会进行熔断，每 n 个函数集成到一起自行定义，会产生很多不必要的工作量和异常。

若多个函数自定义配置相同的 groupKey，但 threadPoolKey 不同，则多个函数将会在不同的线程池中运行。

如果想统一全部 Hystrix 的配置，可在 application 资源配置文件中统一。

如果想统一某个@Service 或@Controller 内的 Hystrix 配置,就需要在@DefaultProperties 注解中进行修饰与配置,或在 application 资源配置文件中用 Hystrix 指向某个@HystrixCommand 进行配置。Hystrix 线程池名称的配置代码如下。

```
@HystrixCommand(
    fallbackMethod="defaultMethod",
    groupKey="indexGroupKey",
    commandKey="indexCommandKey",
```

```
            threadPoolKey="indexThreadPoolKey")
    @RequestMapping(value="/index")
    public Object index() {
        User index = indexService.index();
//      errorService.myError();
        return index;
    }
```

5.3.1　Hystrix 断路器注解式的命令配置

Hystrix 断路器注解式的命令配置主要配置线程名、Hystrix 超时时间、超时检查等。Hystrix 断路器注解式的命令配置伪代码如下。

```
@HystrixCommand(
            fallbackMethod="defaultMethod",
            groupKey="indexGroupKey",
            commandKey="indexCommandKey",
            threadPoolKey="indexThreadPoolKey",
            commandProperties={
                    @HystrixProperty(name ="execution.isolation.thread.timeoutInMilliseconds",value="1000"),
                    @HystrixProperty(name ="execution.timeout.enabled",value="true"),
                    @HystrixProperty(name ="execution.isolation.thread.interruptOnTimeout",value="true"),
                    @HystrixProperty(name ="execution.isolation.semaphore.maxConcurrentRequests",value="10"),
                    @HystrixProperty(name ="fallback.isolation.semaphore.maxConcurrentRequests",value="10"),
                    @HystrixProperty(name ="fallback.enabled",value="true"),
                    @HystrixProperty(name ="circuitBreaker.enabled",value="true"),
                    @HystrixProperty(name ="circuitBreaker.requestVolumeThreshold",value="20")})
    @RequestMapping(value="/index")
    public Object index() {
        User index = indexService.index();
//      errorService.myError();
        return index;
    }
```

@HystrixCommand 注解中使用了 commandProperties 参数，用于编写 Hystrix 的命令内容，即配置 command，commandProperties 参数存储@HystrixProperty 注解的数组。

@HystrixProperty 注解利用 Properties 的键值对形式存储有关 commandProperties 配置的内容。@HystrixCommand 注解中使用@HystrixProperty 注解的 commandProperties 配置参数及释义如表 5-2 所示。

表 5-2　commandProperties 配置的参数及释义

配置的参数及释义	默 认 值
hystrix.command.HystrixCommandKey.execution.isolation.thread.timeoutInMilliseconds： 用于设置命令执行的 timeout 时间，如果发生 timeout，HystrixCommand 会标识为 TIMEOUT，并且执行 Fallback 逻辑；也可以关闭 timeout 检查	1000 毫秒

续表

配置的参数及释义	默 认 值
hystrix.command.HystrixCommandKey.execution.timeout.enabled： 设置是否开启 HystrixCommand.run() 的 timeout 检查	true
hystrix.command.HystrixCommandKey.execution.isolation.thread.interruptOnTimeout： 设置当 HystrixCommand.run() 发生 timeout 后是否需要中断	true
hystrix.command.HystrixCommandKey.execution.isolation.semaphore.maxConcurrentRequests： 设置请求最大的并发数，超过最大并发数的请求将被隔离并拒绝。开启 HystrixCommand.run() 方法后，使用 ExecutionIsolationStrategy.SEMAPHORE 枚举，如果达到最大并发限制，之后的请求将被拒绝。调整信号量时使用的逻辑，与选择向线程池中添加多少线程的逻辑基本相同，但是信号量的开销要小得多，而且执行速度通常要快得多（毫秒级），否则用户将使用线程。例如，单个实例上的 5000 rps 用于查找的内存中被认为只使用了 2 个信号量。隔离原则仍然是一样的，因此信号量仍然应该是整个容器（即 Tomcat）线程池的一小部分，而不是全部或大部分，否则它将不提供保护	10
hystrix.command.HystrixCommandKey.fallback.isolation.semaphore.maxConcurrentRequests： 控制 HystrixCommand.getFallback() 执行。也适用于 ExecutionIsolationStrategy.SEMAPHORE.- fallback.isolation.semaphore.maxConcurrentRequests 和 ExecutionIsolationStrategy.THREAD 属性设置的最大请求数。HystrixCommand.getFallback() 方法允许从调用线程生成。若达到最大并发限制，则随后的请求将被拒绝，并引发异常，因为不能检索回退	10
hystrix.command.HystrixCommandKey.fallback.enabled： 控制是否开启 HystrixCommand.getFallback()	true
hystrix.command.HystrixCommandKey.circuitBreaker.enabled： 控制是否开启熔断器	true
hystrix.command.HystrixCommandKey.circuitBreaker.requestVolumeThreshold： 设置窗口期内触发熔断的最小请求数。例如，将该值设置为 20，当只有 19 个请求时，即使 19 个请求都失败，也不会触发熔断	20

代码如图 5-5 所示。

Hystrix 配置包括命令配置、合并配置和线程池配置，其中命令配置包括执行配置、回退配置、断路配置、度量配置和请求配置。

```
@HystrixCommand(
        fallbackMethod="defaultMethod",
        groupKey="indexGroupKey",
        commandKey="indexCommandKey",
        threadPoolKey="indexThreadPoolKey",
        commandProperties={
                @HystrixProperty(name ="execution.isolation.thread.timeoutInMilliseconds",value="1000"),
                @HystrixProperty(name ="execution.timeout.enabled",value="TRUE"),
                @HystrixProperty(name ="execution.isolation.thread.interruptOnTimeout",value="TRUE"),
                @HystrixProperty(name ="execution.isolation.semaphore.maxConcurrentRequests",value="10"),
                @HystrixProperty(name ="fallback.isolation.semaphore.maxConcurrentRequests",value="10"),
                @HystrixProperty(name ="fallback.enabled",value="TRUE"),
                @HystrixProperty(name ="circuitBreaker.enabled",value="TRUE"),
                @HystrixProperty(name ="circuitBreaker.requestVolumeThreshold",value="20")}
        )
@RequestMapping(value="/index3")
public Object myIndex() {
    User index = indexService.index();
//    errorService.myError();
    return index;
}
```

图 5-5

5.3.2 Hystrix 断路器的注解式线程池配置

前面配置了 Hystrix 的线程名等参数，对 Hystrix 线程池配置的伪代码如下。

```java
@HystrixCommand(
        fallbackMethod="defaultMethod",
        groupKey="indexGroupKey",
        commandKey="indexCommandKey",
        threadPoolKey="indexThreadPoolKey",
        commandProperties={
                @HystrixProperty(name ="execution.isolation.thread.timeoutInMilliseconds",value="1000"),
                @HystrixProperty(name ="execution.timeout.enabled",value="true"),
                @HystrixProperty(name ="execution.isolation.thread.interruptOnTimeout",value="true"),
                @HystrixProperty(name ="execution.isolation.semaphore.maxConcurrentRequests",value="10"),
                @HystrixProperty(name ="fallback.isolation.semaphore.maxConcurrentRequests",value="10"),
                @HystrixProperty(name ="fallback.enabled",value="true"),
                @HystrixProperty(name ="circuitBreaker.enabled",value="true"),
                @HystrixProperty(name ="circuitBreaker.requestVolumeThreshold",value="20")},
        threadPoolProperties= {
                @HystrixProperty(name ="coreSize",value="10"),
                @HystrixProperty(name ="maximumSize",value="10"),
                @HystrixProperty(name ="maxQueueSize",value="-1"),
                @HystrixProperty(name ="queueSizeRejectionThreshold",value="5"),
                @HystrixProperty(name ="keepAliveTimeMinutes",value="1"),
                @HystrixProperty(name ="allowMaximumSizeToDivergeFromCoreSize",value="false"),
        })
@RequestMapping(value="/index")
public Object index() {
    User index = indexService.index();
//    errorService.myError();
    return index;
}
```

threadPoolProperties 参数存储了有关线程池的配置，即对 threadPool 进行配置，内部则需要通过@HystrixProperty 数组进行存储。@HystrixCommand 注解中使用@HystrixProperty 注解的 threadPoolProperties 配置参数及释义如表 5-3 所示。

表 5-3 threadPoolProperties 配置的参数及释义

配置的参数及释义	默 认 值
coreSize： 设置核心线程池大小	10
MaximumSize： 设置最大线程池大小。在不拒绝的情况下支持的最大并发量	10

配置的参数及释义	续表 默 认 值
maxQueueSize： 设置队列大小，-1 为无限大，Hystrix 将使用 SynchronousQueue，若为正值则使用 LinkedBlockingQueue 队列	-1
queueSizeRejectionThreshold： 设置队列的拒绝阈值，但在 maxQueueSize 参数为-1 时不启用	5

通过 Hystrix 的配置内容，可以了解 Hystrix 的功能。代码如图 5-6 所示。

```
@HystrixCommand(
        fallbackMethod="defaultMethod",
        groupKey="indexGroupKey",
        commandKey="indexCommandKey",
        threadPoolKey="indexThreadPoolKey",
        commandProperties={
                @HystrixProperty(name ="execution.isolation.thread.timeoutInMilliseconds",value="1000"),
                @HystrixProperty(name ="execution.timeout.enabled",value="TRUE"),
                @HystrixProperty(name ="execution.isolation.thread.interruptOnTimeout",value="TRUE"),
                @HystrixProperty(name ="execution.isolation.semaphore.maxConcurrentRequests",value="10"),
                @HystrixProperty(name ="fallback.isolation.semaphore.maxConcurrentRequests",value="10"),
                @HystrixProperty(name ="fallback.enabled",value="TRUE"),
                @HystrixProperty(name ="circuitBreaker.enabled",value="TRUE"),
                @HystrixProperty(name ="circuitBreaker.requestVolumeThreshold",value="20")},
        threadPoolProperties= {
                @HystrixProperty(name ="coreSize",value="10"),
                @HystrixProperty(name ="maximumSize",value="10"),
                @HystrixProperty(name ="maxQueueSize",value="-1"),
                @HystrixProperty(name ="queueSizeRejectionThreshold",value="5"),
                @HystrixProperty(name ="keepAliveTimeMinutes",value="1"),
                @HystrixProperty(name ="allowMaximumSizeToDivergeFromCoreSize",value="false")}
)
@RequestMapping(value="/index4")
public Object myIndex() {
        User index = indexService.index();
//      errorService.myError();
        return index;
}
```

图 5-6

5.3.3 Hystrix 断路器注解式的整体定制配置

5.2 节通过 Hystrix 的命令注解@HystrixCommand 进行配置，但如果每个 Hystirx 都按照如图 5-7 所示的方式定制，代码的数量和程序员的工作量都十分庞大。

如果单个@Service 或@Controller 需要进行整体配置，可以使用@DefaultProperties 注解，代码如图 5-7 所示。所有 Hystrix 的配置信息都在 com.netflix.hystrix.contrib.javanica.annotation.DefaultProperties 注解中，@Controller 中每个@HystrixCommand 都会按照所在类的默认配置进行配置。

注意，@DefaultProperties 注解中的回退参数为 defalutFallback，而@HystrixCommand 的回退参数为 fallbackMethod。@DefaultProperties 注解中不含有 commandKey 参数，commandKey 参数相当于@HystrixCommand 的独立名，默认为函数名，默认配置中不会对其进行统一设置。若需单独定制，可写在@HystrixCommand 中，因为@DefaultProperties 不包含 commandKey 属性。

```
package com.zfx.controller;
import org.springframework.beans.factory.annotation.Autowired;
@RestController
@DefaultProperties(
        defaultFallback="defaultMethod",
        groupKey="indexGroupKey",
        threadPoolKey="indexThreadPoolKey",
        commandProperties={
                @HystrixProperty(name ="execution.isolation.thread.timeoutInMilliseconds",value="1000"),
                @HystrixProperty(name ="execution.timeout.enabled",value="true"),
                @HystrixProperty(name ="execution.isolation.thread.interruptOnTimeout",value="true"),
                @HystrixProperty(name ="execution.isolation.semaphore.maxConcurrentRequests",value="10"),
                @HystrixProperty(name ="fallback.isolation.semaphore.maxConcurrentRequests",value="10"),
                @HystrixProperty(name ="fallback.enabled",value="true"),
                @HystrixProperty(name ="circuitBreaker.enabled",value="true"),
                @HystrixProperty(name ="circuitBreaker.requestVolumeThreshold",value="20")},
        threadPoolProperties= {
                @HystrixProperty(name ="coreSize",value="10"),
                @HystrixProperty(name ="maximumSize",value="10"),
                @HystrixProperty(name ="maxQueueSize",value="-1"),
                @HystrixProperty(name ="queueSizeRejectionThreshold",value="5"),
                @HystrixProperty(name ="keepAliveTimeMinutes",value="1"),
                @HystrixProperty(name ="allowMaximumSizeToDivergeFromCoreSize",value="false")
        })
public class IndexController2 {
    @Autowired
    private IndexService indexService;
    @Autowired
    private ErrorService errorService;
    @HystrixCommand
    @RequestMapping(value="/index2")
    public Object index() {
        User index = indexService.index();
//      errorService.myError();
        return index;
    }
    private String defaultMethod(Throwable throwable) {
        System.out.println("已进入private String defaultMethod()");
        System.out.println(throwable.getMessage());
        return "系统超时";
    }
}
```

图 5-7

5.3.4 Hystrix 断路器资源配置式的整体定制配置

5.3.3 节介绍了如果多个@HystrixCommand 在一个@Service 或@Controller 类中，需使用@DefaultProperties 注解进行配置。如果想要配置的@HystrixCommand 不在一个@Service 或@Controller 类中，可以通过配置文件对其进行配置，整体的 application.yml 资源配置文件伪代码如下。

```
hystrix:
  command:
    indexCommandKey:
      execution:
        timeout:
          enabled: true
        isolation:
          thread:
            timeoutInMilliseconds: 3000
            interruptOnTimeout: true
```

```
    semaphore:
        maxConcurrentRequests: 10
```

此处与 5.3.3 节中的配置相同,指定 CommandKey 等于 index 的全部函数都会执行此配置。CommandKey 可以任意编写。

如果将 CommandKey 写成 default,那么没有被定制的@HystrixCommand 都会使用 default 编写的 Hystrix 配置。

5.4 【实例】Hystrix 断路器的请求缓存

5.4.1 实例背景

5.2 节中将异常、超时、高并发进行降级回退处理,但是 Web 项目经常出现重复性高并发的情况。例如,请求某个热销品的售价会一次次请求服务器,获得某 ID 下商品的详情、库存等信息。

Hystrix 给用户提供了相应的缓存能力。Hystrix 请求缓存的原理是在 Hystrix 监听某个函数被调用后,利用"该函数+入参=缓存 Key,返回内容=value"的方式,将缓存内容放置到 Hystrix 的上下文中,待下次获取时直接从 Hystrix 的上下文中获取,而不会再调用其他接口或函数。

本实例复制 5.2 节的 cloud-book-8087-2 工程,称为 cloud-book-8088-3,使用 cloud-user-8083 工程中的一个含参的 Feign Client 调用 cloud-book-8088-3 工程,观察 Hystrix 请求缓存的运行效果。Feign Client 等部分代码不再赘述。

5.4.2 通过 Filter 初始化 Hystrix 上下文

Hystrix 需要利用 Filter 过滤器初始化其上下文,并且每次执行请求时,都会通过过滤器存入 Hystrix 上下文中,Filter 实现类 HystrixFilter.java 代码如下。

```
package com.zfx.filter;
import java.io.IOException;
import javax.servlet.Filter;
import javax.servlet.FilterChain;
import javax.servlet.FilterConfig;
import javax.servlet.ServletException;
import javax.servlet.ServletRequest;
import javax.servlet.ServletResponse;
import javax.servlet.annotation.WebFilter;
import com.netflix.hystrix.strategy.concurrency.HystrixRequestContext;
@WebFilter(urlPatterns="/*",filterName="hystrixFilter",asyncSupported=true)
public class HystrixFilter implements Filter{
    @Override
    public void init(FilterConfig filterConfig) throws ServletException {}
    @Override
```

```
    public void doFilter(ServletRequest request, ServletResponse response, FilterChain chain) throws
IOException, ServletException {
        HystrixRequestContext context = HystrixRequestContext.initializeContext();
        try {
            chain.doFilter(request, response);
        } catch(Exception e){
            e.printStackTrace();
        } finally {
            context.shutdown();
        }
    }
    @Override
    public void destroy(){}
}
```

（1）在上述段代码中增加了@WebFilter 注解，即 javax 包下注解，它将实现 Filter 过滤器的类注入 Spring 容器中，声明 Filter 过滤器，让 Spring Boot 应用程序增加 hystrixFilter 过滤器。@WebFilter 注解配置的参数及释义如表 5-4 所示。

表 5-4 @WebFilter 注解配置的参数及释义

参　　数	释　　义
UrlPatterns	过滤的 URL 路径
filterName	Filter 名称（要求唯一）
asyncSupported	是否进行异步处理（不牵扯 Filter 响应时使用，其阻塞操作会通过不同线程异步执行上下文）

（2）HystrixRequestContext 为 Hystrix 请求上下文，Hystrix 请求的缓存就存储在 Hystrix 上下文中。因此使用 Hystrix 缓存前需要先初始化 Hystrix 请求上下文，并且每次请求时都通过 Hystrix 请求上下文查看是否含有相同 Key 值的 Hystrix 缓存，若 Context 上下文中有，则不调用其他接口，直接从 Context 上下文中获取相应数据并进行缓存。

（3）chain.doFilter(request, response);表示继续执行 Filter 链路中的其他 Filter，待所有 Filter 执行后，释放请求进入相应@Controller 或 Servlet 中。

（4）对于 HystrixRequestContext 上下文，无论在 Feign 的 RequestInterceptor 拦截器中进行初始化加载，还是在 SpringBoot 的 Filter 过滤器中进行初始化加载，都没有任何区别。

（5）@WebFilter 注解中扫描的 urlPatterns 为 "/*"，即在该 Spring Boot HTTP 下，任意@RequestMapping 路径的请求前和请求成功后的返回前，都会进入@WebFilter 注解修饰的 Filter 类 doFilter 函数中。

（6）由于本实例微服务注册在 Consul 注册中心上，并且给 Consul 提供了 Health()函数，每隔一段时间 Consul 都会请求 Health()函数，确认该微服务是否正在正常运行。此时，在 Consul 请求 Health()函数前和 Spring Boot 给 Consul 返回结果前，都会进入@WebFilter 修饰的 Filter 过滤器中。因此通常情况下，urlPatterns 为 "/*" 的 Filter 中建议不输出日志内容，否则无效的重复性日志将过多。

5.4.3 让启动类扫描 Filter 过滤器

从下述代码开始，将使用@SpringCloudApplication 注解，启动类 ApplicationMain-8088.java 代码如下。

```java
package com.zfx;
import org.springframework.boot.SpringApplication;
import org.springframework.boot.web.servlet.ServletComponentScan;
import org.springframework.cloud.client.SpringCloudApplication;
import org.springframework.cloud.netflix.ribbon.RibbonClient;
import org.springframework.cloud.netflix.ribbon.RibbonClients;
import org.springframework.cloud.openfeign.EnableFeignClients;
import org.springframework.web.bind.annotation.RequestMapping;
import org.springframework.web.bind.annotation.RestController;
import com.zfx.config.ribbon.UserRibbonConfig;
@RestController
@EnableFeignClients
@RibbonClients ({
        @RibbonClient(name = "cloud-user", configuration = UserRibbonConfig.class)
})
@SpringCloudApplication
@ServletComponentScan
public class ApplicationMain8088 {
    @RequestMapping("/health")
    public String health() {
        return "200";
    }
    public static void main(String[] args) {
        SpringApplication.run(ApplicationMain8088.class, args);
    }
}
```

通过@ServletComponentScan 注解，扫描 com.zfx 包下是否含有 javax 包中的 Servlet 相关注解，开始使用已声明的@WebFilter 或@WebServlet 实现。如果没有扫描，javax 包注解不会被使用。一般将 lang、util 放在 java 多个包中，servlet 放在 javax 包中，在 REST 风格刚兴起使用 JBOSS 类库时，是通过 javax 包中的@Path、@Get 和@Post 注解实现的，JBOSS 平台会扫描 javax 包中的@Path、@Get 和@Post 注解。

5.4.4 编写 Controller 的 Helper 类

Controller 的 Helper 类是编者习惯的一种写法，就如很多架构要求返回值统一为 Result 对象一样，都是一种编写风格。在实际项目中不应该把逻辑放在@Controller 里，导致@Controller 一个函数要两三千行代码，Helper 可以帮助拆分@Controller 代码。

编写 Helper 类时还需注意 Hystrix 的请求缓存必须在单独的类中。Hystrix 的接口函数和请求缓存函数不能在相同的类中，否则会导致缓存不生效。IndexHelper 类中的代码如下。

```java
package com.zfx.controller.helper;
import java.util.Random;
import org.springframework.beans.factory.annotation.Autowired;
import org.springframework.stereotype.Service;
import com.netflix.hystrix.contrib.javanica.annotation.HystrixCommand;
import com.netflix.hystrix.contrib.javanica.cache.annotation.CacheResult;
import com.netflix.hystrix.strategy.concurrency.HystrixRequestContext;
import com.zfx.entity.User;
import com.zfx.service.IndexService;
@Service
public class IndexHelper {
    @Autowired
    private IndexService indexService;
    @CacheResult(cacheKeyMethod="getUserAge")
    @HystrixCommand(fallbackMethod="defaultMethod")
    public String doSomething(String age) {
        String name = Thread.currentThread().getName();
        System.out.println("当前线程名为： " + name);
        User user = new User();
        user.setAge(Integer.valueOf(age));
        String something = indexService.getUser1(user);
        Random random = new Random();
        System.out.println("something"+something+"||"+random.nextInt());
        return something;
    }
    @SuppressWarnings("unused")
    private String defaultMethod(String age) {
        System.out.println("已进入 private String defaultMethod()");
        return "系统超时";
    }
    public String getUserAge(String age) {
        System.out.println("public String getUserAge(String age) {"+age);
        return age;
    }
}
```

Hystrix 上下文初始化后，由@CacheResult 注解所修饰的函数，将被存入 Hystrix 上下文中。@CacheResult 注解含有@CacheKey 注解，用于辅助设置缓存 Key 名。

getUserAge 函数会输出缓存中 age 的 Value 值。注意，getUserAge 函数必须返回 String 类型，否则会报错。

5.4.5 编写 Controller 类

为方便观察测试效果，编写 IndexController.java 类，将 IndexController.java 中 Hystrix 相关的内容全部放在 Helper 中，此时代码十分简捷。IndexController.java 类代码如下。

```
package com.zfx.controller;
import org.springframework.beans.factory.annotation.Autowired;
import org.springframework.web.bind.annotation.RequestMapping;
import org.springframework.web.bind.annotation.RequestMethod;
import org.springframework.web.bind.annotation.RestController;
import com.zfx.controller.helper.IndexHelper;
import com.zfx.entity.User;
@RestController
public class IndexController {
    @Autowired
    private IndexHelper indexHelper;
    @RequestMapping(value="/index",method = RequestMethod.POST)
    public Object index(User user) {
        String index = null;
        System.out.println("public Object index(User user) {"+user);
        for (int i = 0; i < 10; i++) {
            index = indexHelper.doSomething(String.valueOf(user.getAge()));
        }
        return index;
    }
}
```

在实际项目中，通常在@HystrixCommand 注解修饰的函数内会调用多个@Service，因为 Hystrix 缓存是占用内存的，所以 Hystrix 缓存的目的是整个@Controller 接口的返回，而不是@Service 的缓存。

注意，如果 Helper 中的@CacheResult 直接写在@Controller 中，由其他@RequestMapping 修饰的函数进行调用，那么@CacheResult 注解不会生效，因为@CacheResult 注解也是基于 Spring AOP 编写的。IndexController 中请求了 10 次 doSomething()函数，如果未使用缓存，cloud-user 微服务应该被调用 10 次；而如果使用缓存，cloud-user 微服务只被调用 1 次。

5.4.6 当前项目结构

图 5-8 中的 cloud-book-8088-3 微服务将接收多次请求，并将多个请求的线程合并为一个线程，且只访问一次图 5-9 中的 cloud-user-8083 微服务。

在当前项目结构中，因为 Hystrix 的请求缓存需要通过 Filter 初始化 Hystrix 上下文，所以新增了 HystrixFilter.java 类，但结构其他部分与 5.2 节的项目结构相同。

第 5 章 分布式的断路器

图 5-8   图 5-9

5.4.7 运行结果

（1）在未使用 Hystrix 请求缓存的情况下，Feign Client 端的日志如图 5-10 所示。

```
something{"name":"已进入cloud-book-8083项目getUser当中","age":123123}||1441837310
当前线程名为: hystrix-IndexHelper-2
已进入AdminInterceptor拦截器之中
something{"name":"已进入cloud-book-8083项目getUser当中","age":123123}||-91136216
当前线程名为: hystrix-IndexHelper-3
已进入AdminInterceptor拦截器之中
something{"name":"已进入cloud-book-8083项目getUser当中","age":123123}||-512401078
当前线程名为: hystrix-IndexHelper-4
已进入AdminInterceptor拦截器之中
something{"name":"已进入cloud-book-8083项目getUser当中","age":123123}||-1532672029
当前线程名为: hystrix-IndexHelper-5
已进入AdminInterceptor拦截器之中
something{"name":"已进入cloud-book-8083项目getUser当中","age":123123}||-1583700812
当前线程名为: hystrix-IndexHelper-6
已进入AdminInterceptor拦截器之中
something{"name":"已进入cloud-book-8083项目getUser当中","age":123123}||1975714224
当前线程名为: hystrix-IndexHelper-7
已进入AdminInterceptor拦截器之中
something{"name":"已进入cloud-book-8083项目getUser当中","age":123123}||-27176728
当前线程名为: hystrix-IndexHelper-8
已进入AdminInterceptor拦截器之中
something{"name":"已进入cloud-book-8083项目getUser当中","age":123123}||-1775348493
当前线程名为: hystrix-IndexHelper-9
已进入AdminInterceptor拦截器之中
something{"name":"已进入cloud-book-8083项目getUser当中","age":123123}||-1576229877
当前线程名为: hystrix-IndexHelper-10
已进入AdminInterceptor拦截器之中
something{"name":"已进入cloud-book-8083项目getUser当中","age":123123}||14094427
```

图 5-10

如图 5-10 所示，目前 Feign 调用端使用不同的 thread 线程执行，调用 Feign 提供服务的端口，由于 Hystrix 默认 10 个线程，所以线程名最多只有 10 个。线程的顺序是由 CPU 分配的，不按照数字顺序进行执行也很常见。

（2）在未使用 Hystrix 请求缓存的情况下，Feign Server 端的日志如图 5-11 所示。

```
User [name=null, age=123123]
user.toString()User [name=已进入cloud-book-8083项目getUser当中, age=123123]
User [name=null, age=123123]
user.toString()User [name=已进入cloud-book-8083项目getUser当中, age=123123]
User [name=null, age=123123]
user.toString()User [name=已进入cloud-book-8083项目getUser当中, age=123123]
User [name=null, age=123123]
user.toString()User [name=已进入cloud-book-8083项目getUser当中, age=123123]
User [name=null, age=123123]
user.toString()User [name=已进入cloud-book-8083项目getUser当中, age=123123]
User [name=null, age=123123]
user.toString()User [name=已进入cloud-book-8083项目getUser当中, age=123123]
User [name=null, age=123123]
user.toString()User [name=已进入cloud-book-8083项目getUser当中, age=123123]
User [name=null, age=123123]
user.toString()User [name=已进入cloud-book-8083项目getUser当中, age=123123]
User [name=null, age=123123]
user.toString()User [name=已进入cloud-book-8083项目getUser当中, age=123123]
User [name=null, age=123123]
user.toString()User [name=已进入cloud-book-8083项目getUser当中, age=123123]
```

图 5-11

在未使用 Hystrix 请求缓存的情况下，cloud-user 项目被调用了 10 次。

（3）在使用 Hystrix 请求缓存的情况下，Feign Client 端的日志如图 5-12 所示。

```
已进入Filter拦截器
public Object index(User user) {User [name=123123123, age=123123]
public String getUserAge(String age) {123123
public String getUserAge(String age) {123123
当前线程名为: hystrix-IndexHelper-5
已进入AdminInterceptor拦截器中
something{"name":"已进入cloud-book-8083项目getUser当中","age":123123}||687246909
public String getUserAge(String age) {123123
public String getUserAge(String age) {123123
public String getUserAge(String age) {123123
public String getUserAge(String age) {123123
public String getUserAge(String age) {123123
public String getUserAge(String age) {123123
public String getUserAge(String age) {123123
public String getUserAge(String age) {123123
public String getUserAge(String age) {123123
public String getUserAge(String age) {123123
public String getUserAge(String age) {123123
public String getUserAge(String age) {123123
public String getUserAge(String age) {123123
public String getUserAge(String age) {123123
public String getUserAge(String age) {123123
public String getUserAge(String age) {123123
public String getUserAge(String age) {123123
```

图 5-12

在使用 Hystrix 请求缓存的情况下，1 次调用后直接从上下文中返回，而不去请求 cloud-user。

（4）在使用 Hystrix 请求缓存的情况下，Feign Server 端的日志如图 5-13 所示。

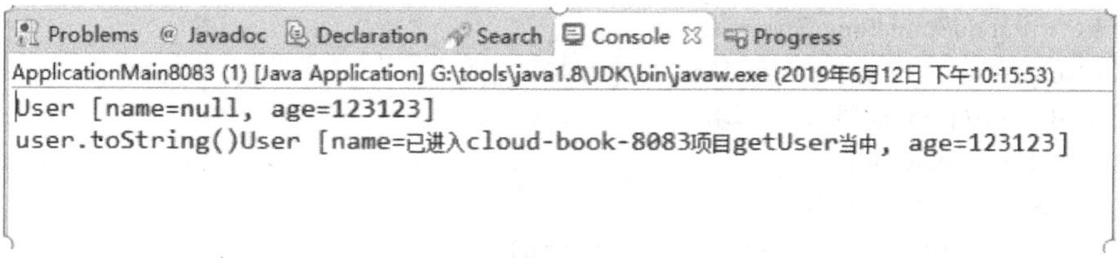

```
Problems  @ Javadoc  Declaration  Search  Console ⊠  Progress
ApplicationMain8083 (1) [Java Application] G:\tools\java1.8\JDK\bin\javaw.exe (2019年6月12日 下午10:15:53)
User [name=null, age=123123]
user.toString()User [name=已进入cloud-book-8083项目getUser当中, age=123123]
```

图 5-13

从 Feign Server 端的日志可以看出，cloud-user 项目只被请求了 1 次。

5.4.8 销毁 Hystrix 的请求缓存

@CacheRemove 注解用来声明 Hystrix 缓存销毁注解，需配合@HystrixCommand 注解使用，在 commandKey 中输入的是需要删除的命令，详解可参考 5.4 节。@CacheKey 注解用于配合@CacheRemove 注解,删除 commandKey = "doSomething"命令下 Key=#{age}的缓存。

在实际项目中 Hystrix 请求缓存的销毁部分代码，可以写在任务调度内，定时调用缓存销毁功能，方便程序未来维护。下述代码的作用是清除 Hystrix 中 doSomething 命令下以 age 作为 Key 的缓存，销毁 Hystrix 请求缓存的伪代码如下。

```
@CacheRemove(commandKey = "doSomething")
@HystrixCommand(fallbackMethod="defaultMethod")
public String removeKey(@CacheKey String age) {
    return "清除成功";
}
```

5.4.9 实例易错点

1．cache 不能放在被调用的类中

Hystrix 的 cache 缓存不能够和调用缓存的函数放置在同一个类中，缓存与调用函数在同一个类时将使缓存失效，该函数被正常调用而不会使用缓存，错误代码如下。

```
@RestController
public class IndexController {
    @Autowired
    private IndexService indexService;
    @RequestMapping(value="/index",method = RequestMethod.POST)
    public Object index(User user) {
        String index = null;
        System.out.println("public Object index(User user) {"+user);
        for (int i = 0; i < 10; i++) {
            index = doSomething(String.valueOf(user.getAge()));
        }
```

```
        return index;
    }
    @CacheResult(cacheKeyMethod="getUserAge")
    @HystrixCommand(fallbackMethod="defaultMethod")
    public String doSomething(String age) {
        String name = Thread.currentThread().getName();
        System.out.println("当前线程名为： " + name);
        User user = new User();
        user.setAge(Integer.valueOf(age));
        String something = indexService.getUser1(user);
        Random random = new Random();
        System.out.println("something"+something+"||"+random.nextInt());
        return something;
    }
}
```

由于@CacheResult 底层是依靠 Spring AOP 编写的，所以上述代码将调用的函数与@CacheResult 修饰的函数放在同一个类中，Hystrix 缓存则可能不会生效。

2．未找到 Hystrix 的 Context 请求上下文

未找到 Hystrix 的 Context 请求上下文的报错信息如下。

java.lang.IllegalStateException: Request caching is not available. Maybe you need to initialize the HystrixRequestContext?

解决方法如下。

（1）先确认是否编写 HystrixRequestContext.initializeContext();初始化函数。

（2）如果在 Filter 中编写了初始化函数，可检查是否使用@WebFilter 将 Filter 进行声明；如果确认已经使用@WebFilter 声明，可检查 Spring Boot 启动类是否使用@ServletComponentScan 扫描@WebFilter 注解，将声明的@WebFilter 注册到 Spring 容器中。

（3）如果在 Feign 的拦截器 RequestInterceptor 中，初始化 HystrixRequestContext 上下文，可检查是否在 RequestInterceptor 拦截器中使用@Component 或@Configuration 等注解，将 RequestInterceptor 拦截器注册到 Spring 容器中。

（4）如果在普通函数中编写初始化函数，可检查是否先进入初始化函数，再调用@CacheResult 修饰的函数。

3．入参变量名不可更改

HystrixRequestContext 上下文中依靠 doSomething(String age)中的 age，作为 Cache 缓存的 Key 进行存储。如果更改了 age 入参，在 HystrixRequestContext 上下文中就无法获得相同 Key 名的缓存，即无法从 HystrixRequestContext 上下文中获取缓存并返回，此时@HystrixCommand 会正常请求 cloud-user 微服务。

4．找不到 CacheKeyMethod 函数

如果自定义 CacheKeyMethod 函数且名字错误，就会报以下错误，即找不到 getUserAge2

函数。此时需检查 com.zfx.controller.helper.IndexHelper 类中是否含有该函数。

> org.springframework.web.util.NestedServletException: Request processing failed; nested exception is com.netflix.hystrix.contrib.javanica.exception.HystrixCachingException: method with name 'getUserAge2' doesn't exist in class 'class com.zfx.controller.helper.IndexHelper'

默认情况下 CacheKeyMethod 函数会自动生成，本实例主要为了运行结果上的展示效果。不编写 CacheKeyMethod 函数的 Helper 代码如下。

```
@Service
public class IndexHelper {
    @Autowired
    private IndexService indexService;
    @CacheResult
    @HystrixCommand(fallbackMethod="defaultMethod")
    public String doSomething(String age) {
        String name = Thread.currentThread().getName();
        System.out.println("当前线程名为： " + name);
        User user = new User();
        user.setAge(Integer.valueOf(age));
        String something = indexService.getUser1(user);
        Random random = new Random();
        System.out.println("something"+something+"||"+random.nextInt());
        return something;
    }
    @SuppressWarnings("unused")
    private String defaultMethod(String age) {
        System.out.println("已进入 private String defaultMethod()");
        return "系统超时";
    }
}
```

上述代码与 5.4.7 节的代码执行结果相同。

5.5 【实例】Hystrix 的请求合并

5.5.1 实例背景

如果某个请求并发多、经过的服务多、调用链长，查询数据库的次数就非常多，占用大量服务器资源，可使用请求合并的方案进行处理。

请求合并会将多个线程合并成一个线程执行，类似于合并线程。Hystrix 请求合并相当于对合并线程进行了整理和实现。如果用户通过商品 ID 调用某一个接口获取数据库中商品的实体类，在 10 秒内需要请求 20 次该接口，但是每次只能传输 1 个商品 ID，此时可使用 Hystrix 请求合并将 20 次调用的 ID 通过 ThreadLocal 线程副本整理到一个 Thread 线程中，请求合并能有效地减少线程数目，减轻服务器的资源压力。

本实例将继续更改 5.4 节的 cloud-book-8090-4 工程与 cloud-user-8083 工程。先删除原来的 Hystrix 请求缓存部分代码，然后在 cloud-book-8090-4 工程中使用 Hystrix 的 @HystrixCollapser 请求合并注解，通过 cloud-book-8090-4 工程多次调用 cloud-user-8083 工程，将多次调用合并到一个线程中调用，最后观察运行结果。

5.5.2　增加 @HystrixCollapser 请求合并修饰的函数

此处直接修改了 5.4.4 节的 cloud-book-8090-4 工程的 IndexHelper.java 类，代码如下。

```java
package com.zfx.controller.helper;
import java.util.ArrayList;
import java.util.List;
import java.util.concurrent.Future;
import org.springframework.beans.factory.annotation.Autowired;
import org.springframework.stereotype.Service;
import com.netflix.hystrix.HystrixCollapser.Scope;
import com.netflix.hystrix.contrib.javanica.annotation.HystrixCollapser;
import com.netflix.hystrix.contrib.javanica.annotation.HystrixCommand;
import com.netflix.hystrix.contrib.javanica.annotation.HystrixProperty;
import com.zfx.entity.User;
import com.zfx.service.IndexService;
@Service
public class IndexHelper {
    @Autowired
    private IndexService indexService;
    @HystrixCollapser(batchMethod = "doSomethings",scope = Scope.GLOBAL, collapserProperties = {
            @HystrixProperty(name="timerDelayInMilliseconds", value = "3000")
    })
    public Future<User> doSomething(String age) {
        System.out.println("****");
        return null;
    }
    @HystrixCommand()
        public List<User> doSomethings(List<String> ages) {
        System.out.println("collapsingListGlobal 当前线程" + Thread.currentThread().getName());
        System.out.println("当前请求参数个数:" + ages.size());
        System.out.println("ages 参数" + ages);
        List<User> users = new ArrayList<>();
        for (String age : ages) {
            User user = new User();
            user.setAge(Integer.valueOf(age));
```

```
                users.add(user);
                String returnUser = indexService.getUser1(user);
                System.out.println(returnUser);
            }
            return users;
        }
    }
```

（1）@HystrixCollapser 注解为请求合并注解，被该注解修饰的函数在被调用后不会直接进入函数内部，而直接跳转到注解的 batchMethod 参数中编写的函数，该函数名可自行定义。@HystrixCollapser 注解配置的参数及释义如表 5-5 所示。

表 5-5 @HystrixCollapser 注解配置的参数及释义

参　　数	释　　义
batchMethod	请求合并后需跳转的函数位置
Scope	请求合并的范围，默认为 Scope.REQUEST（对一次请求的多次服务调用进行合并），可更改为 Scope.GLOBAL（对多次请求的多次服务调用进行合并）
collapserProperties	对@HystrixCollapser 进行配置

（2）@HystrixCollapser 修饰的函数只能使用 Future<?>类型返回。若@Hystrix- Collapser 修饰的函数使用 batchMethod 指向函数的返回类型（List<User>），不会发生任何异常或报错，但也不会进行请求合并，而会多次调用 batchMethod 函数。若@HystrixCollapser 修饰的函数使用除 Future<?>和 batchMethod 的返回值外的返回值，本次请求则会报错。

（3）java.util.concurrent.Future<?>对象是并发编程中的基础对象，Future<?>对象通过 get()函数直接返回泛型中的类型和 Future 存储的泛型中的值。

在 Java 中创建一个线程类，需要继承 Thread 或实现 Runnable，通常情况下如果 Java 函数 A 内部使用了一个自定义的 Thread 线程 B，那么 A 函数执行到 B 线程类后，不会等待 B 线程的返回结果，A 函数的线程与 B 线程类将会使用两个线程同时执行，若 A 函数先执行完整个函数并返回给前台，则不会管理 B 线程类的任何事情。

但如果 A 函数需要等待 B 线程返回结果，B 线程只能选择返回 Future<?>对象，并且在 A 函数执行到 B 线程类后，会形成两个线程执行，但是此刻 A 函数的线程会阻塞，直到 B 线程返回了 Future 结果，A 函数的线程才会继续执行。

（4）@HystrixCollapser 修饰的函数，使用 batchMethod 参数进行请求合并后的调用，batchMethod 指向的函数必须使用@HystrixCommand()注解修饰。

（5）@HystrixCollapser 修饰的函数中 System.out.println("****");不会进行输出，因为在使用 doSomething()函数时，直接跳转到 doSomethings()函数，调用时完全不会进入 doSomething()函数。

（6）由于请求合并后将合并多个 doSomething()函数，所以需要将 doSomething()函数中的入参与出参，转化 doSomething()函数入参与出参集合的形式。

（7）请求合并将调用 doSomething()函数的多个线程合并成一个线程，并且该线程在循环调用 indexService.getUser1()函数时，与正常调用无任何区别。@HystrixCollapser 注解不能管理 indexService.getUser1()函数。

5.5.3 Controller 中调用请求合并函数

因为 Hystrix 请求合并由 Spring AOP 实现，所以调用请求合并的函数与@HystrixCollapser 修饰的请求合并函数不能在同一个类中，如果在相同类中，@HystrixCollapser 修饰的函数会正常执行，但是不能合并线程。

请求合并的 IndexController.java 类代码如下，此 IndexController.java 类中的 index()函数循环 10 次调用 indexHelper.doSomethong()函数。

```java
package com.zfx.controller;
import java.util.concurrent.ExecutionException;
import org.springframework.beans.factory.annotation.Autowired;
import org.springframework.web.bind.annotation.RequestMapping;
import org.springframework.web.bind.annotation.RequestMethod;
import org.springframework.web.bind.annotation.RestController;
import com.zfx.controller.helper.IndexHelper;
import com.zfx.entity.User;
@RestController
public class IndexController {
    @Autowired
    private IndexHelper indexHelper;
    @RequestMapping(value="/index",method = RequestMethod.POST)
    public String index(User user) throws InterruptedException, ExecutionException {
        System.out.println("public Object index(User user) {"+user);
        for (int i = 0; i < 10; i++) {
            System.out.println("public Object index{i"+i);
            indexHelper.doSomething(String.valueOf(user.getAge()+i));
        }
        return "success";
    }
}
```

（1）如果使用返回值，那么以上 10 次请求不会合并到一起，因为需要等待有返回值的线程的返回值并输出，所以线程不会阻塞等待其他线程，耽误线程的返回时间。因此要注意区分书写有无返回值的情况。关于 Future<?>函数与异步线程的内容，将在第 6 章中利用多个 Demo 及运行结果进行介绍。

（2）在调用 indexHelper.doSomething()函数使用@HystrixCollapser 请求合并前，必须初始化 Hystrix 的 HystrixRequestContext 上下文，5.4.2 节中的实例已经使用 Filter 初始化 HystrixRequestContext 上下文。

5.5.4 当前项目结构

图 5-14 中的 cloud-book-8089-4 微服务与图 5-15 中的 cloud-user-8083 微服务，与 3.3 节中的 cloud-user-8083 微服务结构相同。

图 5-14

图 5-15

5.5.5 运行结果

请求合并成功的日志如图 5-16 所示。

```
已进入Filter拦截器
public Object index(User user) {User [name=123, age=123321]
public Object index{i0
public Object index{i1
public Object index{i2
public Object index{i3
public Object index{i4
public Object index{i5
public Object index{i6
public Object index{i7
public Object index{i8
public Object index{i9
collapsingListGlobal当前线程hystrix-IndexHelper-1
当前请求参数个数:10
ages参数[123323, 123326, 123324, 123328, 123325, 123322, 123329, 123327, 123330, 123321]
已进入AdminInterceptor拦截器之中
```

图 5-16

请求合并失效的日志如图 5-17 所示。

```
已进入public class IndexFallBack implements IndexService index
User [name=fall back2 中的user, age=123321]
public Object index{i1
collapsingListGlobal当前线程hystrix-IndexHelper-2
当前请求参数个数:1
ages参数[123322]
已进入AdminInterceptor拦截器之中
2019-06-25 21:25:42.892  WARN 7008 --- [ix-cloud-user-2] com.
已进入public class IndexFallBack implements IndexService index
User [name=fall back2 中的user, age=123322]
```

图 5-17

如图 5-16 所示，程序将会循环 10 次。与请求合并成功的日志最直观的区别是，在请求合并失效的情况下，每次请求 doSomethings 都只能给 List<String> ages 传输 1 个请求参数；而在合并请求成功的情况下，将 10 个参数同时传入 doSomethings 函数的 ages 入参中，即将 10 次请求合并到一起。

5.5.6　实例易错点

1．合并请求不可使用返回值

在下述代码的@Controller 接口中，index()接口调用了 10 次 doSomething()函数，但是使用了 doSomething()的返回值，Future 对象通过 get()函数将 User 返回给 index()函数。

```java
@RequestMapping(value="/index",method = RequestMethod.POST)
public String index(User user) throws InterruptedException, ExecutionException {
    System.out.println("public Object index(User user) {"+user);
    Future<User> doSomething = null;
    List<User> users = new ArrayList<>();
    for (int i = 0; i < 10; i++) {
        System.out.println("public Object index{i"+i);
        doSomething = indexHelper.doSomething(String.valueOf(user.getAge()+i));
        users.add(doSomething.get());
    }
    System.out.println("***"+users);
    return "success";
}
```

上述代码可以执行成功，但是由于每次执行都需要把 doSomething()的返回值存入变量 users 中，导致 doSomethings()无法进行请求合并，只能一次次返回 doSomething()函数的相应结果。

除了 Hystrix 请求合并会导致上述返回值问题，new Thread 独立线程池返回、@Async 异步线程池返回、相关线程池返回等都会导致上述问题。第 6 章将详细介绍@Async 异步线程池及返回值等内容。

2．找不到请求合并参数

找不到请求合并参数时会报如下错误。此时需检查 batch method 函数是否名为 doSomething2，或检查 doSomething2 函数与调用 doSomething2 的函数是否在同一个类中。

```
org.springframework.web.util.NestedServletException: Request processing failed; nested exception is java.lang.IllegalStateException: batch method is absent: doSomethings2
    at org.springframework.web.servlet.FrameworkServlet.processRequest(FrameworkServlet.java:982)
    at org.springframework.web.servlet.FrameworkServlet.doPost(FrameworkServlet.java:877)
    at javax.servlet.http.HttpServlet.service(HttpServlet.java:661)
```

3．合并后入参错误

合并后入参错误的报错信息如下。doSomething()函数的入参类型为 String，

doSomethings()请求合并函数应用 List 包裹 doSomething()函数的入参 String，即 List<String>。若输入其他类型的入参，则报错，如 List<Integer>、String 等。

如果 doSomething()函数的入参类型为 List<String>，doSomethings()请求合并函数可用 List 将 doSomething()函数的入参类型再包裹一层，即 List<List<String>>类型。

org.springframework.web.util.NestedServletException: Request processing failed; nested exception is java.lang.IllegalStateException: required batch method for collapser is absent, wrong generic type: expected com.zfx.controller.helper.IndexHelper.doSomethings(java.util.List<class java.lang.String>), but it's class java.lang.Integer

5.6 【实例】Hystrix 的可视化监控

5.6.1 实例背景

Hystrix 自带可视化监控的仪表盘 dashboard，用于监控@HystrixCommand 注解执行 Hystrix 线程池的使用情况，监控线程池的剩余线程数目、活跃数目、线程执行次数、执行时间等。

本实例使用 4 个微服务整合并观察 Hystrix 的可视化监控，分别为 cloud-user-8083、cloud-admin-8084、cloud-book-8087-2 和 cloud-dashboard-8090。

cloud-dashboard-8090 微服务将为整体项目提供可视化仪表盘的页面，并展示 cloud-book-8087-2 微服务调用 cloud-admin-8084 微服务，cloud-admin-8084 微服务调用 cloud-user-8083 微服务。最后通过可视化仪表盘查看相应结果，不过由于 Hystrix 的可视化监控只能监控一层@HystrixCommand 注解调用，所以看不到 cloud-user-8083 微服务相关的内容。

5.6.2 Hystrix 可视化监控的依赖

cloud-dashboard-8090 微服务与 cloud-book-8087-2 微服务都需要添加依赖，pom.xml 文件所需增加依赖代码如下。

```xml
<dependency>
    <groupId>org.springframework.cloud</groupId>
    <artifactId>spring-cloud-starter-netflix-hystrix-dashboard</artifactId>
</dependency>
```

5.6.3 Hystrix 可视化监控的启动类

cloud-dashboard-8090 微服务统一管理 Hystrix 的可视化仪表盘，只需在 cloud-dashboard-8090 微服务中，增加启动 Hystirx 可视化监控的@EnableHystrixDashboard 注解即可。Application8090.java 启动类代码如下。

```java
package com.zfx;
import org.springframework.boot.SpringApplication;
import org.springframework.boot.web.servlet.ServletComponentScan;
import org.springframework.cloud.client.SpringCloudApplication;
```

```
import org.springframework.cloud.netflix.hystrix.dashboard.EnableHystrixDashboard;
import org.springframework.cloud.netflix.ribbon.RibbonClient;
import org.springframework.cloud.netflix.ribbon.RibbonClients;
import org.springframework.cloud.opcnfcign.EnableFeignClients;
import org.springframework.web.bind.annotation.RequestMapping;
import org.springframework.web.bind.annotation.RestController;
import com.zfx.config.ribbon.UserRibbonConfig;
@RestController
@EnableFeignClients
@RibbonClients ({
        @RibbonClient(name = "cloud-user", configuration = UserRibbonConfig.class)
})
@SpringCloudApplication
@ServletComponentScan
@EnableHystrixDashboard
public class ApplicationMain8090 {
    @RequestMapping("/health")
    public String health() {
        return "200";
    }
    public static void main(String[] args) {
        SpringApplication.run(ApplicationMain8090.class, args);
    }
}
```

为方便日后项目的维护管理工作，统一将@Enable**等相关启动注解都放在启动类中。

5.6.4 被监控的微服务增加响应地址

2.3.9 节提到，Consul 通过 HTTP 不断 ping 微服务的 Health()接口，以保持微服务的健康。Hystrix 响应地址的含义与 Health()健康响应地址类似。

Hystrix 的响应地址需要通过@Bean 注解自定义一个响应地址，并将 Hystrix 的 Servlet 拦截器注册在 Spring 容器中。cloud-dashboard-8090 管理可视化的微服务除不断 ping Hystrix 的 Servlet 外，还会获取 Hystrix 的上下文相关数据，Application8087.java 启动类代码如下：

```
package com.zfx;
import org.springframework.boot.SpringApplication;
import org.springframework.boot.web.servlet.ServletRegistrationBean;
import org.springframework.cloud.client.SpringCloudApplication;
import org.springframework.cloud.netflix.ribbon.RibbonClient;
import org.springframework.cloud.netflix.ribbon.RibbonClients;
import org.springframework.cloud.openfeign.EnableFeignClients;
import org.springframework.context.annotation.Bean;
import org.springframework.web.bind.annotation.RequestMapping;
import org.springframework.web.bind.annotation.RestController;
```

```java
import com.netflix.hystrix.contrib.metrics.eventstream.HystrixMetricsStreamServlet;
import com.zfx.config.ribbon.UserRibbonConfig;
@RestController
@EnableFeignClients
@RibbonClients ({
    @RibbonClient(name = "cloud-admin", configuration = UserRibbonConfig.class),
    @RibbonClient(name = "cloud-user", configuration = UserRibbonConfig.class)
})
@SpringCloudApplication
public class ApplicationMain8087 {
    @RequestMapping("/health")
    public String health() {
        return "200";
    }
    @Bean
    public ServletRegistrationBean<HystrixMetricsStreamServlet> hystrixMetricsStreamServlet() {
        ServletRegistrationBean<HystrixMetricsStreamServlet> registration
            = new ServletRegistrationBean<HystrixMetricsStreamServlet>(new HystrixMetricsStreamServlet());
        registration.addUrlMappings("/hystrix.stream");
        return registration;
    }
    public static void main(String[] args) {
        SpringApplication.run(ApplicationMain8087.class, args);
    }
}
```

被可视化管理的 cloud-book-8087-2 微服务需编写上述代码，给予前台可调用的 Hystrix Servlet 地址。注意，cloud-book-8087-2 微服务不需要开启@EnableHystrixDashboard 可视化管理页面，只需提供地址即可。

被可视化管理的 cloud-book-8087-2 微服务除在 Spring 容器中进行注册@Bean 外，还需要在 application.yml 文件中配置 Hystirx 框架，声明 cloud-dashboard-8090 微服务需通过该@Bean 提供的 Servlet 进行调用，application.yml 配置文件代码更改如下。

```yaml
management:
  endpoints:
    web:
      exposure:
        include: hystrix.stream
```

5.6.5 当前项目结构

cloud-user-8083、cloud-admin-8084、cloud-book-8087-2 等相关微服务与前文中的项目结构相同。cloud-dashboard-8090 微服务本身也只需要一个启动类与一个 application.yml 资源配置文件，此处略过。

5.6.6　运行结果

先启动所有的微服务 cloud-user-8083、cloud-admin-8084、cloud-book-8087-2 和 cloud-dashboard-8090，如图 5-18 所示。

图 5-18

打开 Hystrix 的可视化监控页面，地址为 localhost:8087/hystrix，如图 5-19 所示。

图 5-19

turbine.stream：需输入 5.6.4 节中的响应地址，即 http://被监控微服务的 IP + 被监控微服务的 HOST/被监控的响应地址（UrlMappings）。

Delay：需输入刷新间隔，建议设置长间隔时间，因为 Hystrix 的上下文比较多，若间隔短则经常会发生卡死的现象，默认为 2000 毫秒。

Title：需输入任意字符串，无太多意义，只作为仪表盘的标题使用。

输入相关内容后，通过地址 http://localhost:8087/hystrix.stream，可查看 Hystirx 的 dashboard 仪表盘，如图 5-20 所示。

此仪表盘需@HystrixCommand 注解被正常调用后才可显示，如果被调用后还显示为空，需等待一定时间进行刷新。

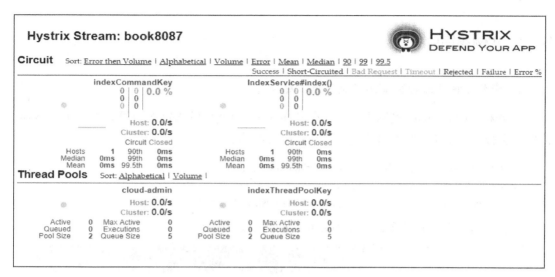

图 5-20

如果直接通过 URL 调用 http://localhost:8087/hystrix.stream，页面如图 5-21 所示。

图 5-21

hystrix.stream 拦截器返回的是 Hystrix 相关的上下文，dashboard 的功能是将以上数据进行可视化处理。注意，hystrix.stream 拦截器除了返回 Hystrix 上下文，还有 ping 的功能。

5.6.7 实例易错点

1. 未在 Client 端配置调用地址或地址错误

如果未在 Client 端配置调用地址或在 Dashboard 页面处输入错误的调用地址，Dashboard 微服务会报以下 INFO 级别日志进行警告，即非异常通知。

```
2019-06-26 23:07:15.141    INFO 4468 --- [io-8090-exec-10]
ashboardConfiguration$ProxyStreamServlet :
Proxy opening connection to: http://localhost:8087/hystrix.stream
```

前台页面为空，并不会显示其他异常，如图 5-22 所示。

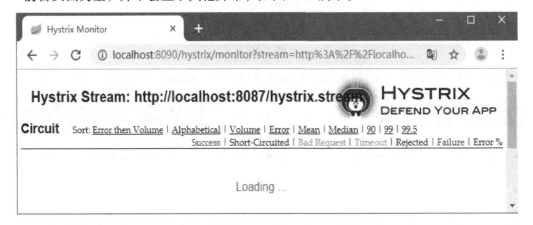

图 5-22

2. 未在 Dashboard 微服务开启监控页面功能

未通过注解在启动类开启 Hystrix 可视化监控，前台会报 404 错误，并找不到 Hystrix 相关管理页面，如图 5-23 所示。

图 5-23

5.7 本章小结

5.2 节实现了更高级的降级回退方式，在一个函数中使用多个 Feign Client 的 Service，若有任何异常，整个函数都会回退到降级函数中。

5.4 节实现了在调用某个函数后，将该函数的返回结果作为缓存，以防多次调用产生高并发，用"函数名+入参"的形式作为 Key 值，将返回结果作为 Value 值缓存。在一个函数中使用多个 Feign Client 的 Service，以减小多个微服务之间沟通的并发压力。

5.5 节实现了在多次调用某个函数后，将多个线程合并成一个线程进行调用，减小了系统内存和线程并发数量的压力。在一个函数中使用某个 Feign Client 的 Service，将多次得到的 ID 合并成一个 List 对其他微服务进行请求，减小多个微服务之间沟通的并发压力。

5.6 节利用 Hystrix 自带的 Dashboard 性能监控仪表盘页面和性能监控控件，监控其他微服务的 Hystrix 线程池，包括 Hystrix 线程池目前的剩余线程数目、线程池容量、并发情况、运行情况、执行次数等。其实 Hystrix 自带的 Dashboard 仪表盘使用方法十分简单，在需要被监控的微服务中通过 application 资源配置文件打开被调用地址，通过@Bean 注解将性能控件相关内容配置在 Spring 容器上，监控端 Dashboard 仪表盘使用@EnableHystrixDashboard 注解开启页面，并在页面上调用 Hystrix 性能控件的相关地址即可。

分布式的三大剑客：注册中心 Consul（包括通信 Feign）、微服务 Spring Application、断路器 Hystrix 已经全部介绍完成。一般在分布式系统搭建初期，先要搭建三个基本要素，再在每个微服务中增加增删改查等业务逻辑。

5.8 习 题

1. 创建 Spring Boot 工程 1 与 Spring Boot 工程 2，使其注册在 Consul 注册中心集群上，使用 Feign 让 Spring Boot 工程 1 与 Spring Boot 工程 2 进行通信。整合 Hystrix 保证降级回退，强制 Spring Boot 工程 2 进行报错，记录程序状态。

2. 创建 Spring Boot 工程 1、Spring Boot 工程 2 与 Spring Boot 工程 3，Spring Boot 工程 1 调用 Spring Boot 工程 2，由 Spring Boot 工程 2 调用 10 次 Spring Boot 工程 3，观察程序在使用 Hystrix 请求缓存和未使用请求缓存的不同情况，记录调用次数等信息。若在使用 Hystrix 请求缓存途中，Spring Boot 工程 3 突然关机，会发生什么情况。若 Hystrix 使用请求合并（不使用请求缓存），会发生什么情况。若同时使用请求合并和请求缓存，会发生什么情况。

3. 整合 Hystrix 可视化监控，在前 2 个习题运行过程中，记录 Hystrix 可视化监控响应的信息。

第 6 章　微服务的异步线程池

5.8 节通过 Hystrix 将请求的线程进行合并，简单介绍了 Future 线程返回类，本章详细介绍线程池与线程池返回之间的时间与关系。

6.1　异步线程池

在代码自上而下执行的过程中，若其中有一个动作是异步的，则可以不用关心异步线程中所处理的逻辑，代码会继续执行，但是异步线程池中的代码会单独执行。通常情况下，某同步函数中 new Thread().start 运行的独立线程对于同步函数而言都是异步线程。若不考虑返回值，则主线程不会等待异步线程的执行时间，直接往下执行。

Java 代码大多是通过同步的方式进行处理的，但是与第三方进行交互时，如短信、邮件、支付等，因网络或第三方执行时间容易造成响应时间较长，而应用程序会暂停并等待第三方的回复。因此自 Spring 3.0 版本开始内置了 @Async，以解决无须强一致性结果却长时间等待问题。

6.1.1　异步线程池特点

异步线程池的代码书写简单，只需要一个注解即可进行操作。初始化配置已被设置完成。异步线程池会减小程序现在进行时间上的压力，使用户操作时无须等待。异步线程池对服务器性能有要求，虽然时间上的压力减小了，但是所执行的线程数目会变得更多。因此不要频繁使用异步线程池。

6.1.2　常见的线程池

表 6-1 所示为 Java 常见的线程池，Java 的线程池通常使用 ExecutorService 工厂类进行创建，java.util.concurrent.ExecutorService 工厂类下可以调用 java.util.concurrent.ExecutorService.newFixedThreadPool() 等线程池。每个实现的线程池特性都不同，所以配合的场景也不同。

除了 Java 自带的线程池，也有很多第三方线程池可供使用。异步线程池可以通过 ExecutorService 工厂指定使用其中某一个线程池。

表 6-1　Java 常见的线程池

newCachedThreadPool();	可缓存线程池。若线程池长度超过处理需要，可灵活回收空闲线程；若无可回收，则新建线程
newFixedThreadPool();	指定工作线程数量的线程池。每提交一个任务就创建一个工作线程，若工作线程数量达到线程池初始的最大数，则将提交的任务存入线程池队列中
newSingleThreadExecutor();	一个单线程化的 Executor，只创建唯一的工作者线程来执行任务，保证所有任务按照指定顺序（FIFO, LIFO, 优先级）执行。如果该线程异常结束，会有另一个取代它，保证顺序执行。单线程最大的特点是可保证顺序执行每个任务，并且在任意给定的时间内不会有多个线程是活动的
newScheduleThreadPool();	一个定长的线程池，且支持定时和周期性任务执行

6.2　【实例】创建无返回值异步线程池

6.2.1　实例背景

假设在运行一个程序后，同步线程中间进行调用接口或插入一张表等操作需要 3 秒，为了更加快速地反映给前台，使用异步线程池可减少调用接口的 3 秒。同步线程中的其他操作则不需要耽误此时间。调用接口后直接会得到返回值，其他操作会在后台自己进行。

本实例将创建 boot_03_1 工程，利用 @Async 注解和 @EnableAsync 注解完成上述场景的代码编写，初步认识异步线程池及其使用方法。

6.2.2　编写 Pom 文件

从以下 pom.xml 文件代码中可以看出，本实例只使用了普通的微服务工程，没有集成分布式相关内容。在此多依赖了一个 devtools 工程，原因是在前面章节中如果想更新代码，必须先关闭正在启动的服务，然后对代码进行更新，再次运行项目需重新启动；加上依赖后，则不需要关闭启动的服务，直接在 Eclipse 中更新代码并保存，正在运行的服务器会自动进行热部署、部分功能重启、重启 Spring 容器等操作。此依赖包只需要在 pom.xml 中引入即可，不需要编写任何配置文件。

```xml
<parent>
    <groupId> org.springframework.boot </groupId>
    <artifactId> spring-boot-starter-parent </artifactId>
    <version>1.5.3.RELEASE</version>
</parent>
<dependencies>
    <dependency>
        <groupId> org.springframework.boot </groupId>
        <artifactId> spring-boot-starter-web </artifactId>
    </dependency>
    <!-- 添加热部署，开发者模式 -->
    <dependency>
```

```xml
        <groupId>org.springframework.boot</groupId>
        <artifactId>spring-boot-devtools</artifactId>
        <optional>true</optional>
    </dependency>
</dependencies>
<!-- 作为可执行 Jar 包 -->
<build>
    <plugins>
        <plugin>
            <groupId> org.springframework.boot </groupId>
            <artifactId> spring-boot-maven-plugin</artifactId>
        </plugin>
    </plugins>
</build>
```

6.2.3 编写 Spring Boot 启动类

注意，在 main 方法的类名头上要添加@EnableAsync 注解，否则异步线程池不会生效，虽然不会报错，但是当调用时系统依旧同步进行而非异步。@EnableAsync 注解的作用是开启异步线程池。启动类 ApplicationMain.java 代码如下。

```java
package com.zfx;
import org.springframework.boot.SpringApplication;
import org.springframework.boot.autoconfigure.SpringBootApplication;
@EnableAsync
@SpringBootApplication
public class ApplicationMain {
    public staticvoid main(String[] args) {
        SpringApplication.run(ApplicationMain.class,args);
    }
}
```

6.2.4 编写异步线程池任务接口与实现

Service 的 interface 接口只定义了一个无返回值的函数，无返回值异步线程池接口 GatewayService.java 代码如下。

```java
package com.zfx.service;
public interface GatewayService {
    void getAlipayGateway(String username);
}
```

通过实现 GatewayService 接口，可以在实现类中输出相关日志、线程停留时间等。无返回值异步线程池实现类 GatewayServiceImpl.java 代码如下。

```java
package com.zfx.service.impl;
import org.springframework.scheduling.annotation.Async;
import org.springframework.stereotype.Service;
```

```
import com.zfx.service.gatewayService;
@Service
public class GatewayServiceImpl implements gatewayService {
    @Async
    @Override
    public void getAlipayGateway(String username) {
        System.out.println("已进来获得网关路径服务, username 为: " + username + "线程名为: " +Thread.currentThread().getName());
        Long startTime = System.currentTimeMillis();
        try {
            Thread.sleep(5000);
        } catch (InterruptedException e) {
            e.printStackTrace();
        }
        Long endTime = System.currentTimeMillis();
        System.out.println("线程: "+Thread.currentThread().getName()+" 消耗的时间为: "+(endTime-startTime) +"ms");
    }
}
```

@Async 注解将函数标记为"异步"执行的候选函数,也可以用于类级别,如果用于类级别,该类中的所有函数都被认为是异步的。修饰的返回类型被限制为 Void 或 Future。

因为 Future 对象可以追踪异步函数的结果,所以通过 Future 对象可使同步线程获得异步线程的返回结果。

6.2.5 编写外部可调用接口

为方便测试,编写 gatewayController.java 接口类,用一个普通的@Controller 接口循环调用 5 次 Service, gatewayController.java 代码如下。

```
package com.zfx.controller;
import java.util.concurrent.ExecutionException;
import org.springframework.beans.factory.annotation.Autowired;
import org.springframework.web.bind.annotation.PathVariable;
import org.springframework.web.bind.annotation.RequestMapping;
import org.springframework.web.bind.annotation.RestController;
import com.zfx.service.gatewayService;
@RestController
public class gatewayController {
    @Autowired
    gatewayService gatewayService;
    @RequestMapping("/gateway/{username}")
    public String gateway(@PathVariable final String username) throws InterruptedException, ExecutionException{
        Long startTime = System.currentTimeMillis();
        for (int i = 0; i< 5; i++) {
            gatewayService.getAlipaygateway(username);
```

```
        }
        Long endTime = System.currentTimeMillis();
        System.out.println("调用 gateway 消耗的时间为: "+(endTime-startTime)+"ms");
        return "调用 gateway 消耗的时间为: "+(endTime-startTime) +"ms";
    }
}
```

6.2.6 当前项目结构

图 6-1 所示为 boot_03_1 工程目录结构，此处没有将调用方和被调用的@Async 放在同一个类中，因为@Async 底层也是由 Spring AOP 进行管理的。

6.2.7 运行程序查看异步线程池效果

运行程序时查看异步线程池的日志，如图 6-2 所示。虽然执行 gateway()函数需要 5 秒等待时间，但是由于 gateway()函数是异步的，所以无论执行多少次 gateway()函数，接口的返回时间都无限接近于 0 毫秒，前台需要 3 毫秒就能得到接口的返回结果。

图 6-1

图 6-2

代码中循环使用了 Service 中的一个函数，函数的目的是停留 5 秒，模拟调用其他网关的情况。循环了 5 次，正常应该停留 25 秒才能在前台页面返回结果，但事实上，刚刚调用接口后大约 1 秒就返回目前的时间给前台页面，而后台服务器一直继续运行线程，即互相是异步进行的，没有任何时间上的影响。

注意，此刻的线程没有任何先后顺序，每次运行会根据 CPU 的情况不同，线程的执行顺序不同，如下所示。

```
已进来获得网关路径服务, username 为: zhangfangxing 线程名为: SimpleAsyncTaskExecutor-1
已进来获得网关路径服务, username 为: zhangfangxing 线程名为: SimpleAsyncTaskExecutor-2
已进来获得网关路径服务, username 为: zhangfangxing 线程名为: SimpleAsyncTaskExecutor-3
调用 gateway 消耗的时间为: 3ms
已进来获得网关路径服务, username 为: zhangfangxing 线程名为: SimpleAsyncTaskExecutor-4
已进来获得网关路径服务, username 为: zhangfangxing 线程名为: SimpleAsyncTaskExecutor-5
线程: SimpleAsyncTaskExecutor-2 消耗的时间为: 5000ms
线程: SimpleAsyncTaskExecutor-1 消耗的时间为: 5000ms
线程: SimpleAsyncTaskExecutor-3 消耗的时间为: 5000ms
```

线程：SimpleAsyncTaskExecutor-4 消耗的时间为：5000ms
线程：SimpleAsyncTaskExecutor-5 消耗的时间为：5000ms

6.2.8 实例易错点

注意，创建无返回值异步线程池只要使用了@EnableAsync 注解和@Async 注解，通常不会报错。

6.3 【实例】创建有返回值异步线程池

6.3.1 实例背景

6.2 节的实例是没有返回值的异步线程池，但有时程序需要异步线程池给程序带来返回值。本实例将创建有返回值异步线程池。

如果单线程执行一个任务需要 10 秒，其中 5 秒进行异步处理，那么整个异步线程池会节省 5 秒，即需要返回的总时间为 5 秒。

原函数："单线程时间"=10 秒。

有异步线程池的函数："异步线程池 5 秒"＋"后续任务 5 秒"－"异步线程与主线程同时执行任务 5 秒"=5 秒。

本实例将修改 6.2 节的 boot_03_2 工程，完成以上任务。

6.3.2 增加新的服务接口

Future<String>是异步线程池可接收的返回参数，在 GatewayService.java 接口中添加有返回值的异步线程池接口伪代码如下。

```
Future<String> getAlipaygatewayString(String username);
```

6.3.3 增加新的服务实现

为创建有返回值的异步线程，只需增加一个实现函数即可，不需要更改其他内容。在 GatewayServiceImpl.java 实现类中添加有返回值的异步线程池伪代码如下。

```
@Async
@Override
public Future<String> getAlipaygatewayString(String username) {
    Future<String>future;
    System.out.println("已进来获得网关路径服务，username 为："+username);
    Long startTime = System.currentTimeMillis();
    try {
        Thread.sleep(5000);
    } catch (InterruptedException e) {
        future = new AsyncResult<String>("错误提示:" + e.getMessage());
        e.printStackTrace();
        return future;
```

```
        }
        Long endTime = System.currentTimeMillis();
        System.out.println("线程: "+Thread.currentThread().getName()+" 消耗的时间为: "+(endTime-startTime)
+"ms");
        future = new AsyncResult<String>("线程: "+Thread.currentThread().getName()+" 消耗的时间为: 
"+(endTime-startTime)+"ms");
        return future;
    }
```

6.3.4 增加新的调用

先执行异步线程池，该异步线程池会停留 5 秒，然后主线程也停留 5 秒，最后使用异步线程池的返回值进行返回，查看调用 gateway 消耗的时间。gatewayController.java 接口类增加新的调用伪代码如下。

```
@RequestMapping("/gateway2/{username}")
public String gateway2(@PathVariable final String username) throws InterruptedException, ExecutionException{
    Long startTime = System.currentTimeMillis();
    //异步去调用线程（5 秒）
    Future<String>alipaygateway = gatewayService.getAlipaygatewayString(username);
    //主线程等待 5 秒
    try {
        Thread.sleep(5000);
    } catch (InterruptedException e) {
        e.printStackTrace();
    }
    Long endTime = System.currentTimeMillis();
    //如果同步则应该是 10 秒，由于是异步，所以耗时应该为 5 秒
    System.out.println("调用 gateway 消耗的时间为: "+(endTime-startTime));
    return alipaygateway.get();
}
```

6.3.5 当前项目结构

图 6-3 所示的工程与图 6-1 所示的工程项目结构相同。

6.3.6 运行程序查看异步线程池效果

在单线程同步的情况下，原来返回给前台需要 10 秒。但是代码中异步线程池停留了 5 秒，模拟后续代码还会运行 5 秒，整个 Controller 被返回给前台时，一共用了 5 秒，而不是 10 秒，说明此时异步线程池已生效。

接口返回效果如图 6-4 所示。username 为

图 6-3

zhangfangxing，调用 gateway 消耗的时间为 5000 毫秒，线程 Simple-AsyncTaskExecutor-2 消耗的时间为 5000 毫秒。

图 6-4

6.3.7 实例易错点

@Async 异步线程池的返回值只能使用 Future 代理类和 AsyncResult 代理类，若使用其他类型则可能报错，或者不会使用异步执行函数，而使用同步执行函数。

6.4 【实例】优化异步线程池

6.4.1 实例背景

优化异步线程池的主要目的是不能因为数据量过大而启动过多的线程，导致服务器宕机，或者因为数据量过小，启动的核心线程数目过少，导致性能缺失。本实例手动将异步线程池@Async 注解指定给某线程池的实现类，并对其进行配置和优化，在调用异步线程池时，实际上是配置后的指定线程池在使用异步线程。

6.4.2 创建初始化线程池配置类

创建配置类 ExecutorConfig.java，需使用@Configuration 注解将其注册在 Spring 容器上。初始化线程池配置类 ExecutorConfig.java 代码如下。

```
package com.zfx.config;
import java.util.concurrent.Executor;
import java.util.concurrent.ThreadPoolExecutor;
import org.springframework.context.annotation.Bean;
import org.springframework.context.annotation.Configuration;
import org.springframework.scheduling.annotation.EnableAsync;
import org.springframework.scheduling.concurrent.ThreadPoolTaskExecutor;
@Configuration
@EnableAsync
public class ExecutorConfig {
    @Bean
    public Executor myExecutor() {
        //新建 ThreadPoolTaskExecutor 作为异步线程池的指定线程池
        ThreadPoolTaskExecutor executor = new ThreadPoolTaskExecutor();
        //配置核心线程数
```

```
        executor.setCorePoolSize(5);
        //配置最大线程数
        executor.setMaxPoolSize(6);
        //配置队列大小
        executor.setQueueCapacity(99999);
        //配置线程池中的线程名称前缀
        executor.setThreadNamePrefix("async-service-");
        // 指定线程策略为 CallerRunsPolicy 拒绝策略
        executor.setRejectedExecutionHandler(new ThreadPoolExecutor.CallerRunsPolicy());
        //执行初始化
        executor.initialize();
        return executor;
    }
}
```

在上述代码中，指定异步线程池为 ThreadPoolTaskExecutor 线程池，并对其进行配置优化。ThreadPoolTaskExecutor 线程池的构造参数及释义如表 6-2 所示。

表 6-2　ThreadPoolTaskExecutor 线程池的构造参数及释义

构造参数	释　义
corePoolSize	线程池维护线程的最小数量，也称核心线程数量
maxPoolSize	线程池维护线程的最大数量，也称最大线程数量
keepAliveTime	线程池维护线程所允许的空闲时间
unit	线程池维护线程所允许的空闲时间的单位
threadNamePrefix	线程池中线程的名称前缀
rejectedExecutionHandler	线程池中指定的处理策略
QueueCapacity	线程池中的缓冲队列

ThreadPoolTaskExecutor 线程池的特性如下。

（1）ThreadPoolTaskExecutor 核心线程数量 corePoolSize 大于当前线程池中线程的数量，即使当前线程池中的线程都处于空闲状态，也要创建足够的线程进行等待，尽可能使当前线程池中线程数量大于核心线程数量。

（2）ThreadPoolTaskExecutor 核心线程数量 corePoolSize 等于当前线程池中线程的数量，在当前缓冲队列 QueueCapacity 未满的情况下，任务会被放入缓冲队列，不会新建线程。

（3）ThreadPoolTaskExecutor 核心线程数量 corePoolSize 小于当前线程池中线程的数量，在当前缓冲队列 QueueCapacity 已满，当前线程池中线程的数量小于 maxPoolSize 最大线程数量的情况下，会创建新的线程执行任务。

（4）ThreadPoolTaskExecutor 核心线程数量 corePoolSize 小于当前线程池中线程的数量，在当前缓冲队列 QueueCapacity 已满，当前线程池中线程的数量等于 maxPoolSize 最大线程数量的情况下，通过 handler 所指定的策略处理此任务。

（5）ThreadPoolTaskExecutor 核心线程数量 corePoolSize 小于当前线程池中线程的数量，在某线程空闲时间超过了 keepAliveTime，线程则被快速失败。

6.4.3 更改无返回值的异步线程池 Service 实现类

更改异步线程池的 Service 实现类，需给异步线程池增加名称，调用异步线程池时即可指向刚刚配置的 Bean 名称，其代码如下。

```
@Async("myExecutor")
@Override
publicvoid getAlipaygateway(String username) {
    System.out.println("已进来获得网关路径服务，username 为："+ username + "线程名为："+ Thread.currentThread().getName());
    Long startTime = System.currentTimeMillis();
    try {
        Thread.sleep(5000);
    } catch (InterruptedException e) {
        e.printStackTrace();
    }
    Long endTime = System.currentTimeMillis();
    System.out.println("线程："+Thread.currentThread().getName()+" 消耗的时间为："+(endTime-startTime) +"ms");
}
```

@Async 注解只能配置其中的 Value 名称，Value 名称代表@Async 将通过 Spring 容器找到指定异步操作的线程池 Bean，并使用该线程池执行函数。

6.4.4 运行程序查看异步线程池效果

注意，在 6.4.2 节中配置核心线程数代码和配置最大线程数代码如下。为了观察异步线程池的初始化效果，可将 Controller 中的循环从 5 次改成 6 次。

```
//配置核心线程数
executor.setCorePoolSize(5);
//配置最大线程数
executor.setMaxPoolSize(6);
```

如图 6-5 所示，返回前台的时间为 0 秒，说明异步线程池没有任何等待即刻直接返回给前台。核心线程数是 5 个，所以在运行程序时会同时启动 5 个线程；而系统循环了 6 次，所以可以看到名为 async-service-4 的线程启动了 2 次，共计使用了 5 个线程，执行了 6 次，如图 6-6 所示。

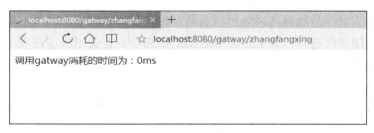

图 6-5

```
Problems  @ Javadoc  Declaration  Console ⊠  Progress
ApplicationMain (20) [Java Application] F:\java1.8\JDK\bin\javaw.exe (2018年9月2日 下午10:19:41)
线程: async-service-5 消耗的时间为: 5000ms
已进来获得网关路径服务, username为: zhangfangxing线程名为: async-service-2
已进来获得网关路径服务, username为: zhangfangxing线程名为: async-service-4
调用gatway消耗的时间为: 0ms
已进来获得网关路径服务, username为: zhangfangxing线程名为: async-service-3
已进来获得网关路径服务, username为: zhangfangxing线程名为: async-service-5
线程: async-service-1 消耗的时间为: 5000ms
已进来获得网关路径服务, username为: zhangfangxing线程名为: async-service-1
线程: async-service-4 消耗的时间为: 5001ms
已进来获得网关路径服务, username为: zhangfangxing线程名为: async-service-4
线程: async-service-5 消耗的时间为: 5001ms
线程: async-service-2 消耗的时间为: 5001ms
线程: async-service-3 消耗的时间为: 5001ms
线程: async-service-1 消耗的时间为: 5000ms
线程: async-service-4 消耗的时间为: 5001ms
```

图 6-6

6.4.5 实例易错点

（1）异步线程池注册在 Spring 容器上的名字，要与@Async()注解内部指定的名字相同，否则可能名称对应不上。

（2）在实例化线程池的过程中，根据线程池的不同，所使用的函数和配置参数不同。最初只要根据不同文档进行搭建即可，后续可以使用 Jmeter 等相关工具对应用程序进行性能测试，根据性能测试的结果对线程池进行调整。

6.5 【实例】优雅停止异步线程池

6.5.1 实例背景

在用@Async 编写优雅关闭的函数时，根据实现类的不同，优雅关闭的方式也不同。本实例将创建 boot_03_4 工程，通过使用 ThreadPoolTaskExecutor 线程池优雅关闭正在执行的实现类，并展示运行效果。

6.5.2 何为"优雅"

了解 Java Socket、Netty、Mina 等技术的人可能会经常看到"优雅"这个词。每个线程在运转时，都需要一定的时间来处理目前的工作任务。但事实上，每次停止服务器，通常会"暴力"关闭系统进程，如 Linux 中的 kill -9 命令，在关闭当前进程时，并没有考虑应用程序内的线程正在处理什么事情，所以 Tomcat 给用户提供了脚本，shutdown.sh 用来优雅停止 Tomcat 服务器。优雅停止是指此刻 Tomcat 不会产生新的线程，同时将目前所正在运行的线程执行完。因此开发 Socket、WebSocket 等重度依赖线程的 Java 程序，需要将线程"优雅"停止，以保证用户在操作时不会遭受任何损失。

6.5.3 修改原 Config 配置类

原 ExcutorConfig.java 异步线程池配置类需要进行如下修改。

（1）增加了一个异步线程池的关闭函数 asyncShutDown(ThreadPoolTaskExecutor executor)，传入的线程池类型即之前创建 myExecutor 异步线程池的线程类型。

（2）executor.setAwaitTerminationSeconds(6);设置最大等待时间为 6 秒，如果 6 秒后异步线程池没有销毁就强制销毁，以确保应用最后能被关闭，而不是阻塞。

（3）executor.setWaitForTasksToCompleteOnShutdown(true);设置异步线程池等待所有任务都结束时，销毁异步线程池。

修改后的 ExecutorConfig.java 异步线程池配置类代码如下。

```java
package com.zfx.config;
import java.util.concurrent.Executor;
import java.util.concurrent.ThreadPoolExecutor;
import org.springframework.context.annotation.Bean;
import org.springframework.context.annotation.Configuration;
import org.springframework.scheduling.annotation.EnableAsync;
import org.springframework.scheduling.concurrent.ThreadPoolTaskExecutor;
@Configuration
@EnableAsync
publicclass ExecutorConfig {

    @Bean(value="myExecutor")
    public Executor myExecutor() {

        ThreadPoolTaskExecutor executor = new ThreadPoolTaskExecutor();
        //配置核心线程数
        executor.setCorePoolSize(5);
        //配置最大线程数
        executor.setMaxPoolSize(6);
        //配置队列大小
        executor.setQueueCapacity(99999);
        //配置线程池中的线程名称前缀
        executor.setThreadNamePrefix("async-service-");
        // rejection-policy: 当 pool 已经达到 max size 时，如何处理新任务
        // CALLER_RUNS: 不在新线程中执行任务，而由调用者所在的线程执行
        executor.setRejectedExecutionHandler(new ThreadPoolExecutor.CallerRunsPolicy());

        //执行初始化
        executor.initialize();
        return executor;
    }

    public boolean asyncShutDown(ThreadPoolTaskExecutor executor) {
        try {
            intpoolSize1 = executor.getPoolSize();
```

```
            System.out.println("当前异步线程池中的线程池最大线程数为: "+poolSize1);
                executor.setWaitForTasksToCompleteOnShutdown(true);
                executor.setAwaitTerminationSeconds(6);
                executor.shutdown();
                System.out.println("当前异步线程池中的线程池最大线程数为: "+poolSize2);
                return true;
            } catch (Exception e) {
                System.out.println("异步线程池关闭出错");
                return false;
            }
        }
    }
```

6.5.4 修改原 Controller 控制层

原 gatewayController.java 接口类需要进行如下修改。

（1）直接将 Spring 容器 ApplicationContext 注入代码中，方便以后直接从容器中获取已注册的异步线程池。

（2）新增一个 shutDown 接口，用来优雅关闭异步线程池。

（3）将刚写的 config 函数注入接口中，从容器中获取已注册的异步线程池进行停机操作，直接返回是否正常关闭异步线程池即可。

修改后的 gatewayController.java 接口类代码如下。

```
package com.zfx.controller;
import java.util.concurrent.ExecutionException;
import java.util.concurrent.Future;
import org.springframework.beans.factory.annotation.Autowired;
import org.springframework.context.ApplicationContext;
import org.springframework.scheduling.concurrent.ThreadPoolTaskExecutor;
import org.springframework.web.bind.annotation.PathVariable;
import org.springframework.web.bind.annotation.RequestMapping;
import org.springframework.web.bind.annotation.RestController;
import com.zfx.config.ExecutorConfig;
import com.zfx.service.gatewayService;
@RestController
Public class gatewayController {

    @Autowired
    private gatewayService gatewayService;
    @Autowired
    private ExecutorConfig executorConfig;
    @Autowired
    private ApplicationContext context;

    @RequestMapping("/gateway/{username}")
    public String gateway(@PathVariable final String username) throws InterruptedException, ExecutionException{
        System.out.println("进入 gateway 接口: "+username);
        Long startTime = System.currentTimeMillis();
```

```java
    for (int i = 0; i < 5; i++) {
        gatewayService.getAlipaygateway(username);
    }
    Long endTime = System.currentTimeMillis();
    System.out.println("调用 gateway 消耗的时间为: "+(endTime-startTime)+"ms");
    return "调用 gateway 消耗的时间为: "+(endTime-startTime) +"ms";
}

@RequestMapping("/gateway2/{username}")
public String gateway2(@PathVariable final String username) throws InterruptedException, ExecutionException{
    System.out.println("进入 gateway2 接口: "+username);
    Long startTime = System.currentTimeMillis();
    //异步调用线程（5秒）
    Future<String> alipaygateway = gatewayService.getAlipaygatewayString(username);
    Long endTime = System.currentTimeMillis();
    System.out.println("调用 gateway2 消耗的时间为: "+(endTime-startTime));
    return alipaygateway.get();
}

@RequestMapping("/shutDown")
public boolean shutDown(){
    System.out.println("调用异步线程池 shutDown 接口");
    //从 Spring 容器中获取线程池
    ThreadPoolTaskExecutor t = (ThreadPoolTaskExecutor)context.getBean("myExecutor");
    boolean asyncShutDown = executorConfig.asyncShutDown(t);
    return asyncShutDown;
}
```

6.5.5 当前项目结构

当前项目结构如图 6-7 所示。

图 6-7

6.5.6 优雅停止异步线程池的执行效果

1. 首先请求 gateway 接口

无返回值执行 6 次的异步线程池接口效果如图 6-8 所示。与之前测试过的相同，前台 2 毫秒即可获得结果，而让异步线程池在后台单独运行。

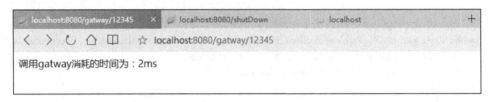

图 6-8

2. 在 gateway 接口未结束时迅速使用 shutDown 接口

效果如图 6-9 所示。shutDown 直接可以返回关闭成功或失败的信息。后续有关线程自身所需的内容及时间，线程池自动运行，无须处理。

图 6-9

3. 在使用 gateway 接口后使用 gateway2 接口

如图 6-10 所示，gateway2 接口一直处于请求状态。因为代码中的逻辑是，先异步停止 5 秒，然后主线程停止 5 秒，但之前用 shutDown 函数将异步线程池优雅停止了，停止后异步线程池是空的，因此无法产生新的线程处理返回结果，线程会一直处于阻塞状态。

图 6-10

4. 日志

如图 6-11 所示，在日志中可以清晰看到代码的执行步骤。

（1）在异步线程池生成 5 个线程对代码进行异步处理。

（2）在还没有进行代码处理时，先调用了 shutDown 接口，优雅停止了异步线程池。

（3）即使优雅停止了异步线程池，也不妨碍之前新产生的 5 个线程进行异步操作，虽然可以在控制台中看到控制台打印出目前异步线程池中含有的 5 个线程后，程序马上调用了异步非阻塞的 shutting down ExecutorService 'myExecutor'，但接下来执行是没有发生任何改变的。

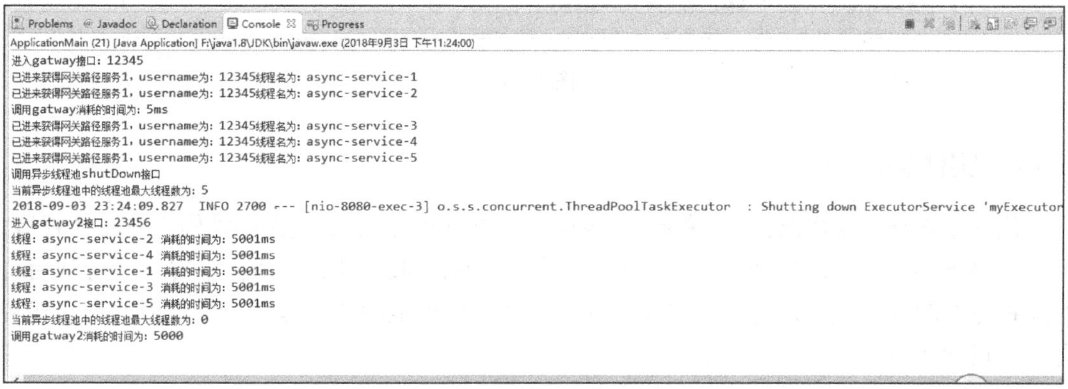

图 6-11

（4）虽然优雅停止了异步线程池，但可以看到 gateway2 接口依旧进入了。说明异步线程池的优雅停止，并不阻止主线程的任何操作。

（5）接下来执行 5 个异步线程后，发现控制台又打印了一次当前线程数，因为在 shutDown 前只执行了 5 个线程，所以等 5 个线程完美执行完成后，异步线程池才会正式停止。此处使用了 ThreadPoolTaskExecutor 的一个技巧，即 shutting down ExecutorService 'myExecutor' 函数本身是异步非阻塞执行的，它会异步等待所有线程执行完成，最终自动关闭。

（6）最后调用 gateway2 消耗的时间为 5000 毫秒，控制台打印说明 gateway2 接口的同步时间已经完成，但是由于当前异步线程池中的线程池最大线程数为 0，所以没有异步线程池去给 gateway2 执行，前台 gateway 会不断请求。

5. 重新启动微服务日志

如图 6-12 所示，前台客户在请求相关地址时，重启微服务，可以发现异步线程池正在运行时会输出异步线程池实现类 serviceImpl 中的相关打印信息，此时才正式进入 gateway2 的异步线程。最后 gateway2 的结果如图 6-13 所示，总调用时间为 5000 毫秒。

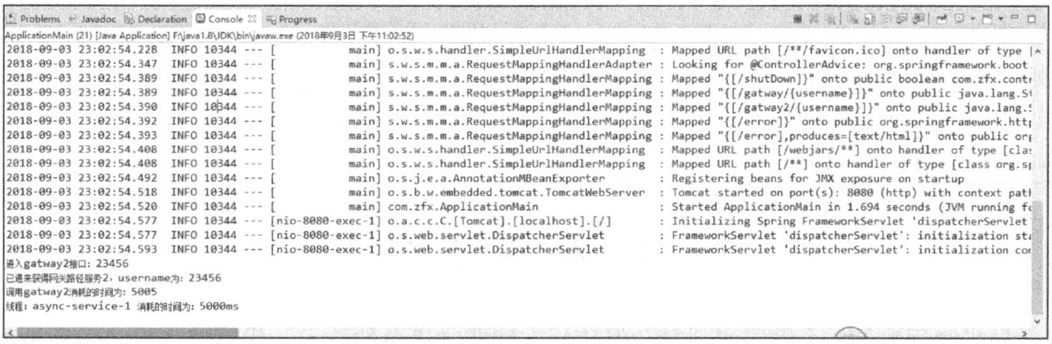

图 6-12

图 6-13

6.5.7 实例易错点

优雅停止异步线程池，不代表停止 Web 服务器，正常情况下应先优雅停止异步线程池，再优雅停止 Web 服务器。但在本实例中不可停止 Web 服务器，优雅停止异步线程池后查看线程池中的线程数目是否正确归零，并观察如果线程池中的线程数目归零后，异步线程有没有正常执行。

6.6 @Enable*注解

如表 6-3 所示，@Enable*相关注解的目的都是使 SpringBoot 微服务项目开启相应功能的启动注解，并无特殊意义。如果需要异步方法支持，就在启动类上增加@EnableAsync 注解；如果还需要计划任务，就在启动类上再增加@EnableScheduling 注解。

@Enable*相关注解类似于一个总开关。若外部注解没有增加@EnableAsync 注解，就相当于整个项目没有异步支持，在没有异步支持（或者异步支持关闭）的情况下，即使内部增加了异步注解，内部的异步注解也不会生效，该函数虽然增加了@Async 注解，在没有发生编译报错的情况下，依旧是同步运行的函数。

表 6-3 @Enable*部分注解及释义

注 解	释 义
@EnableAspectJAutoProxy	开启对 AspectJ 自动代理的支持
@EnableAsync	开启异步方法支持
@EnableScheduling	开启计划任务/任务调度
@EnableWebMvc	开启 Web MVC 配置功能
@EnableHystrix	开启 Hystrix 断路器功能
@EnableConfigurationProperties	开启对@ConfigurationProperties 注解配置 Bean 的支持
@EnableTransactionManagement	开启声明式事务的支持
@EnableCaching	开启注解式的缓存支持

6.7 本章小结

异步线程池是一个底层实现复杂但使用方法非常简单的技术，优雅停止异步线程池的测试执行过程则比较复杂。此处要注意每个细节，尤其是在优雅停止异步线程池且执行完成所有应该执行的线程后，线程池中的线程数目是否正确归零。

如果想了解更多关于优化异步线程池的内容，可参考 Java 自带的线程池、Java 第三方线程池、Java 并发编程等相关内容。不论用 Java 自带的线程池，还是第三方开源的线程池，对于异步线程池来说都只是不同的实现而已，并无区别。因为每种线程池的优雅停止实现都不相同，所以每种不同实现出来的异步线程池的优雅停止也不同。

6.8 习　　题

1. 创建 Spring Boot 工程，整合@Async 异步线程池，在接口处停留 5 秒，异步线程池被调用 5 次，每次停留 5 秒。

（1）记录最后停留时间。

（2）中途强制关闭 Spring Boot 工程，发生何种情况。

（3）中途强制关闭 Spring Boot 工程，保证前台页面无刷新无变化的情况下，迅速重新打开 Spring Boot 工程，发生何种情况。

（4）返回值为 null，记录最后停留时间。

（5）返回值为异步线程池中的相关数据，记录最后停留时间。

（6）返回值为非异步线程池中的相关数据，记录最后停留时间。

2. 创建 Spring Boot 工程，整合@Async 创建无返回值异步线程池，优化异步线程池的相关参数，测试前台调用 shutDown()接口，与强制关闭 Spring Boot 工程是否相同。

3. 创建 Spring Boot 工程，整合@Async 创建无返回值异步线程池，优化异步线程池的相关参数，测试前台调用 gateway1()接口后，调用 shutDown()接口，在后台进程中，调用 gateway2()接口，程序有何反应。注意，gateway2()用的是与 gateway1()相同的线程池或不同的线程池有何不同结果。

4. 创建 Spring Boot 工程，整合@Async 创建无返回值异步线程池，优化异步线程池的相关参数，测试前台调用 gateway1()接口后，停止工程，保证前台页面无刷新无变化的情况下，迅速重新启动工程，程序有何反应。注意，gateway2()用的是与 gateway1()相同的线程池或不同的线程池两种情况有何不同结果。

第 7 章　微服务整合持久化数据源

经过前几章的介绍，已经可以初步搭建一套 Spring Cloud + Spring Boot 的分布式架构程序，主要包括分布式的注册中心、分布式的通信、分布式的断路器。虽然此时还没有深入分布式其他内容，但在分布式架构初具雏形时，可先使微服务整合数据源，以方便日后分布式程序的不断完善。

在分布式架构初具雏形时开始整合数据源，不是为了项目上线赶进度，主要是分布式的鉴权、分布式的任务调度都依赖于数据源，而分布式的消息总线、分布式的链路监控不一定会用到，所以此时集成数据源是在实际工作中最好的选择。

7.1　spring-data

Spring Boot 推荐使用 spring-data 系列产品进行数据库管理和实现。spring-data 系列产品提供一种熟悉的、一致的、基于 Spring 的数据访问编程模型，同时仍然保留底层数据存储的特性。spring-data 使数据访问技术、关系和非关系数据库及基于云的数据服务变得容易，spring-data 是一个包含很多特定数据库子项目的父类项目。

spring-data-jpa 属于 spring-data 数据家族，可以很容易地实现基于 JPA 库的增删改查，主要处理基于 JPA 的数据访问层的增强支持。spring-data-jpa 使构建数据访问技术的 Spring 应用程序变得更加容易。spring-data-jpa 是基于 ORM 规范的解决方案，目的是能够最大程度减轻 Java 程序员操作数据库时的各类代码复杂度。尤其是 spring-data-jpa 提倡的 "约定优于配置" 原则，方便后续对项目维护。

7.1.1　ORM 规范

对象关系映射（Object Relational Mapping，ORM）模式是为了解决面向对象与关系数据库存在的互不匹配现象的技术。ORM 通过使用描述对象和数据库之间映射的元数据，将程序中的对象自动持久化到关系数据库中。

在业务逻辑层和用户界面层中，当对象信息发生变化时，用户需要把对象信息保存在关系数据库中。

ORM 类框架减轻了对数据库控制的压力。只需要有相关的文件映射或类似 SQL 的语句，即可完成对数据库的 CURD 操作。ORM 是一种规范，ORM 类框架是实现 ORM 规范的技术。Hibernate、JPA、My Batis 都属于 ORM 类框架。

7.1.2 JPA、Hibernate、spring-data-jpa 之间的关系

Hibernate 是一套操作数据库的 Java 类框架,其底层封装的 JDBC,根据 ORM 的思想进行制作。因其存在配置映射文件繁多与复杂等弊端,用处越来越少。

JPA 全称为 Java Persistence API,是指通过各种 XML 或注解描述(Java Bean 映射数据库表)的一种实现方案,可以理解为一个抽象的接口或抽象的类。JPA 的底层可以直接调用 Hibernate JPA,宗旨是为 POJO 提供持久化标准规范,能够脱离容器独立运行,方便开发和测试。JPA 通过 JDK 5.0 注解或 XML 描述对象与关系表的映射关系,并将运行期的实体对象持久化到数据库中。

spring-data-jpa 是 Spring 提供的一套更加简捷的 CURD 类框架,其底层封装并简化了 Hibernate 和 JPA 的技术。

7.1.3 安装 MySQL

在 Linux 系统中安装 MySQL 的步骤如下。

安装 MySQL 客户端:

> yum install mysql

安装 MySQL 服务端:

> yum install mysql-server

运行 MySQL 服务端:

> service mysqld start

修改 MySQL 字符集,在/etc/my.cnf 中增加下述代码并重新启动:

default-character-set=utf8

创建 MySQL 的 root 账户:

> mysqladmin -u root password mypassword

登录 MySQL 的 root 账户:

> mysql -u root -p

进入 MySQL 后,新建相关账户:

CREATE USER 'zhang'@'192.168.138.140' IDENTIFIED BY 'mypassword';
#在固定连接地址 192.168.138.140 下创建 zhang 账户,并设置密码为 mypassword

为相关账户授权:

GRANT privileges ON databasename.tablename TO zhang@192.168.138.140
#部分表授权给 zhang 账户,databasename 输入库名,tablename 输入表名,只授权一部分

GRANT all privileges on *.* to zhang@'%'identified by 'mypassword' with grant option;
#全部授权

新建相关账户并授权后进行刷新权限：

FLUSH PRIVILEGES;

安装完后会删除 MySQL 的 test 数据库，以免遭受攻击。配置相关内容后即可在外部使用可视化工具进行管理，如图 7-1 所示。

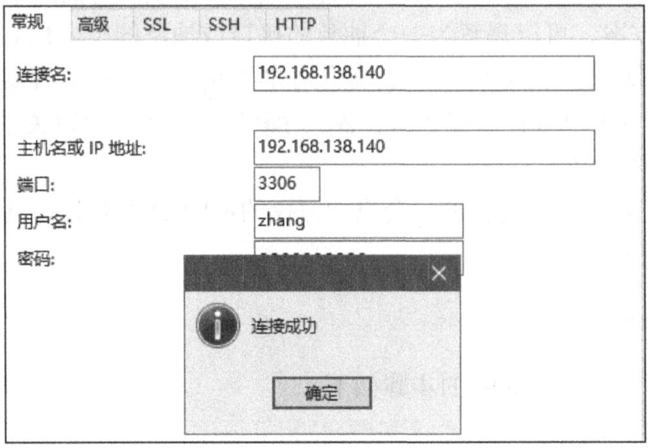

图 7-1

7.2 【实例】Spring Boot 整合 MyBaits 注解式编程

7.2.1 实例背景

MyBaits 注解大约出现在 2010 年，现在大多项目都使用 spring-data-jpa 注解进行整合持久化，两种注解的思想基本相同。本实例先整合 MyBaits 注解，方便后续理解 spring-data-jpa 注解。

本实例将创建 boot_06_4 项目，利用@Mapper、@Select、@Update、@Results、@Result、@Param、@Insert、@Delete、@ConfigurationProperties 等相关 MyBaits 注解进行持久化操作，实现 Java 应用程序对 account 表的增删改查操作。

7.2.2 添加 Pom 文件

Spring Boot 整合 mybatis(xml)、MyBaits 注解都需要依赖 org.mybatis.spring.boot 工程。该工程是 MyBatis 配合 Spring Boot 推出的扩展性工程。对于非注解式的编程方式，本书不再介绍。pom.xml 文件增加相关依赖代码如下。

```
<dependency>
    <groupId>org.mybatis.spring.boot</groupId>
    <artifactId>mybatis-spring-boot-starter</artifactId>
    <version>1.1.1</version>
</dependency>
```

7.2.3 编写 application 资源配置文件

资源配置文件中只需要编写数据库的基本信息即可，application.properties 资源配置文件添加代码如下。

```
spring.datasource.url=jdbc:mysql://192.168.102.128:3306/justLearn
spring.datasource.username=root
spring.datasource.password=root
spring.datasource.driver-class-name=com.mysql.jdbc.Driver
```

7.2.4 编写 dao 层

编写 AccountDao.java 类代码如下。AccountDao.java 类是 MyBaits 注解的核心操作 dao 类，用 interface 接口以@Mapper 注解的方式使程序动态生成 impl 实现类，为用户提供 CRUD 的功能。

```java
package com.zfx.dao;
import java.util.List;
import org.apache.ibatis.annotations.Delete;
import org.apache.ibatis.annotations.Insert;
import org.apache.ibatis.annotations.Mapper;
import org.apache.ibatis.annotations.Param;
import org.apache.ibatis.annotations.Result;
import org.apache.ibatis.annotations.Results;
import org.apache.ibatis.annotations.Select;
import org.apache.ibatis.annotations.Update;
import com.zfx.bean.Account;
import com.zfx.bean.MessageVO;
@Mapper
public interface AccountDao {

    @Insert("insert into 'account' ('account_id', 'password') VALUES (#{accountid}, #{password})")
    public int addAccount(Account Account);

    @Delete("delete from 'account' where account_id=#{accountid}")
    public int deleteByaccountid(String accountid);

    @Update("update 'account' set 'account_id'='#{account.accountid}', 'password'=#{account.password} where 'account_id'=#{account.accountid}")
    public int updateAccount(@Param("account") Account account);

    @Select("select * from 'account' where account_id=#{accountid2}")
    @Results({
      @Result(id=true,property="accountid",column="account_id"),
      @Result(property="password",column="password")
    })
```

```
        public Account queryByAccountid(@Param("accountid2") String accountid);

        @Select("select account_id,message from 'message','account' where account_id=#{accountid2}")
        @Results({
            @Result(property="accountid",column="account_id"),
            @Result(property="message",column="message"),
            @Result(property="message",column="message")
        })
        public List<MessageVO> queryMessageByAccountid(@Param("accountid2") String accountid);

        @Select("select * from 'account'")
        @Results({
            @Result(id=true,property="accountid",column="account_id"),
            @Result(property="password",column="password"),
        })
        public List<Account> findAll();
}
```

@Select 注解可以自动地根据表中的列名与数据库中的名字进行匹配。若其列名相同，则代码可以直接写成如下形式，省略@Results、@Result 注解及其代码，运行后得到的结果相同。

```
@Select("select * from 'account'")
public List<Account> findAll();
```

@Results、@Result 注解主要为了在映射返回值与 Java Bean 参数不相同时，手动指定其参数，代码如下。

```
@Select("select * from 'account'")
@Results({
        @Result(id=true,property="accountid",column="account_id"),
        @Result(property="password",column="password"),
})
public List<Account> findAll();
```

注意，此处将 column 列中名为 account_id 的值，用 property 参数指定给 Account 类的 accountid 参数。

@Param 注解主要为了在映射入参值与 Java Bean 参数不相同时，手动指定其参数，若不需要映射，则代码如下。

```
@Select("select * from 'account' where account_id=#{accountid}")
public Account queryByAccountid(String accountid);
```

若需映射，则代码如下。

```
@Select("select * from 'account' where account_id=#{accountid2}")
@Results({
        @Result(id=true,property="accountid",column="account_id"),
```

```
        @Result(property="password",column="password"),
})
public Account queryByAccountid(@Param("accountid2") String accountid);//注意其写法映射位置
```

将上述代码中的@Param 参数改成了 accountid2，是为了方便介绍所处位置映射的地点，在正常生产中应尽可能保证命名一致。

SQL 语句是以@Param 注解的 Value 参数来进行手动匹配的。在外部调用时直接将要更改的 Java Bean 放入即可，MyBatis 会根据入参自动分配，匹配某 Java Bean 参数的伪代码如下。

```
@Update("update 'account' set 'account_id'='#{account.accountid}', 'password'=#{account.password} where 'account_id'=#{account.accountid}")
public int updateAccount(@Param("account") Account account);
```

@Add 注解只为了使用 insert into 语句而添加。

@Update 注解在进行更改数据库时使用，若发现没有该账号，则新增一个账号；若已有该账号，则改变该账号。

@Mapper 注解在运行程序时，可以初始化 dao 实现类，将其映射成 MyBatis 的 XML 文件。

7.2.5 编写访问接口

为方便测试效果，编写 AccountController.java 类为前台提供 REST 访问接口，其@Controller 接口类或@Service 服务类都可以直接通过@Autowried 注入 AccountDao，即可直接使用 dao 层代码。AccountController.java 类代码如下。

```
package com.zfx.controller;
import java.util.List;
import org.springframework.beans.factory.annotation.Autowired;
import org.springframework.web.bind.annotation.PathVariable;
import org.springframework.web.bind.annotation.RequestMapping;
import org.springframework.web.bind.annotation.RestController;
import com.zfx.bean.Account;
import com.zfx.bean.MessageVO;
import com.zfx.dao.AccountDao;
@RestController
public class AccountController {

    @Autowired
    private AccountDao accountDao;

    @RequestMapping({"/","/index"})
    public List<Account> findAllAccount(){
        return accountDao.findAll();
    }
    @RequestMapping("/query/{accountid}")
```

```java
public Account queryByAccountid(@PathVariable(value="accountid") String accountid){
    return accountDao.queryByAccountid(accountid);
}
@RequestMapping("/update/{accountid}/{password}")
public int updateAccount(@PathVariable String accountid,@PathVariable String password){
    Account account = new Account();
    account.setaccountid(accountid);
    account.setPassword(password);
    return accountDao.updateAccount(account);
}
@RequestMapping("/delete/{accountid}")
public int deleteAccount(@PathVariable String accountid){
    return accountDao.deleteByaccountid(accountid);
}
@RequestMapping("/add/{accountid}/{password}")
public int addAccount(@PathVariable String accountid,@PathVariable String password){
    Account account = new Account();
    account.setaccountid(accountid);
    account.setPassword(password);
    return accountDao.addAccount(account);
}
@RequestMapping("/queryMessage/{accountid}")
public List<MessageVO> queryMessage(@PathVariable String accountid){
    return accountDao.queryMessageByAccountid(accountid);
}
}
```

7.2.6　当前项目结构

如图 7-2 所示，在当前项目结构中只含有上述 4 个 Java 文件。

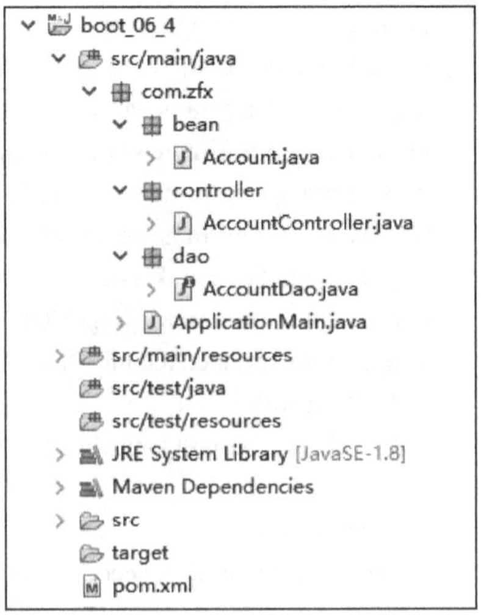

7.2.7　运行效果

只调用 update 接口，任意输入 account_id 和 password 后，若该数据原 account_id 的数据不存在于数据库，则在数据库中新建一条数据，并赋予其相应的信息；若已有 account_id 的数据，则修改其密码。

图 7-2

7.2.8　实例易错点

1．未从数据库得到任何数据包

如果数据库本身没有运行或宕机，会报如下异常，此时应查看 MySQL 是否正常运行。

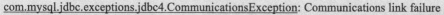

com.mysql.jdbc.exceptions.jdbc4.CommunicationsException: Communications link failure
The last packet sent successfully to the server was 0 milliseconds ago. The driver has not received any packets from the server.

2．数据库连接超时

数据库连接超时可能是因为防火墙挡住了连接，或 MySQL 没有正常运行，此时需通过客户端连接数据库，检查是否能连接成功，并检查数据库是否含有防火墙或没有正常运行现象，异常如下。

Can't connect to MySQL server on '192.168.138.141'(10038)

3．如果没有映射的表

如果没有找到映射的表，会报如下异常，此时应检查数据库中是否含有编写的表，或检查程序中编写的表名是否有错。

\### Error querying database. Cause: com.mysql.jdbc.exceptions.jdbc4.MySQLSyntaxErrorException: Table 'justLearn.account' doesn't exist
\### The error may exist in com/zfx/dao/AccountDao.java (best guess)
\### The error may involve defaultParameterMap
\### The error occurred while setting parameters
\### SQL: select * from 'account'
\### Cause: com.mysql.jdbc.exceptions.jdbc4.MySQLSyntaxErrorException: Table 'justLearn.account' doesn't exist; bad SQL grammar []; nested exception is com.mysql.jdbc.exceptions.jdbc4.MySQLSyntaxErrorException: Table 'justLearn.account' doesn't exist] with root cause com.mysql.jdbc.exceptions.jdbc4.MySQLSyntaxErrorException: Table 'justLearn.account' doesn't exist

4．SQL 语句编写报错

如果@Select 语句中的 SQL 编写报错，会报如下异常，此时需检查报错中提示的语句。

\### Cause: com.mysql.jdbc.exceptions.jdbc4.MySQLSyntaxErrorException: You have an error in your SQL syntax; check the manual that corresponds to your MySQL server version for the right syntax to use near '123select * from 'account" at line 1; bad SQL grammar []; nested exception is com.mysql.jdbc.exceptions.jdbc4.MySQLSyntaxErrorException: You have an error in your SQL syntax; check the manual that corresponds to your MySQL server version for the right syntax to use near '123select * from 'account" at line 1] with root cause com.mysql.jdbc.exceptions.jdbc4.MySQLSyntaxErrorException: You have an error in your SQL syntax; check the manual that corresponds to your MySQL server version for the right syntax to use near '123select * from 'account" at line 1

5．没有输入密码

如果没有输入密码，会报如下异常，注意程序中连接的并不是 192.168.138.1 地址，而是 192.168.138.147 地址。在没有输入密码的情况下，会导致信息载入不全。

注意，如果数据库没有设置密码，在微服务中也无法输入密码，也会报如下错误，MyBaits 注解和 spring-data-jpa 都可能出现此问题。此时应当给数据库设置一个密码。

java.sql.SQLException: Access denied for user 'root'@'192.168.138.1' (using password: NO)

6．密码错误

Druid 连接池报告连接 MySQL 的密码错误，异常如下，此时需检查密码配置。

```
com.alibaba.druid.pool.DruidDataSource    : create connection SQLException, url: jdbc:mysql://192.168.138.141:3306/justLearn, errorCode 1045, state 28000
```

7.3 @Mapper 注解详解

7.3.1 @Mapper 和 XML 形式的对应关系

用前文代码举例如下，介绍@Mapper 注解与 XML 文件的对应关系。

```
@Select("select * from 'account' where account_id=#{accountid}")
@Results({
    @Result(id=true,property="accountid",column="account_id"),
    @Result(property="password",column="password"),
})
public Account queryByAccountid(@Param("accountid") String accountid);

<select id="queryByAccountid" resultMap="resultMap1">
    select * from 'account' where account_id=#{accountid,jdbcType=VARCHAR}
</select>
<resultMap type="com.zfx.bean.Account" id="resultMap1">
    <id column="account_id" property="accountid">
    <result conlumn="password" property="password">
</resultMap>
```

上述代码体现了@Mapper 注解和 XML 写法相对应的情况，两者意义相同，执行内容相同。只是@Mapper 注解会自动匹配并实现 IMPL。

7.3.2 MyBatis 的注解式编程多表查询

在数据库中增加 message 表，如图 7-3 所示，将 acountid 和 account 表进行关联。

查询某账户 ID 可以获得用户的全部消息和该用户发表的信息，进行多表查询的代码如下。注意，此处 SQL 语句使用了 message 和 account 两张表。

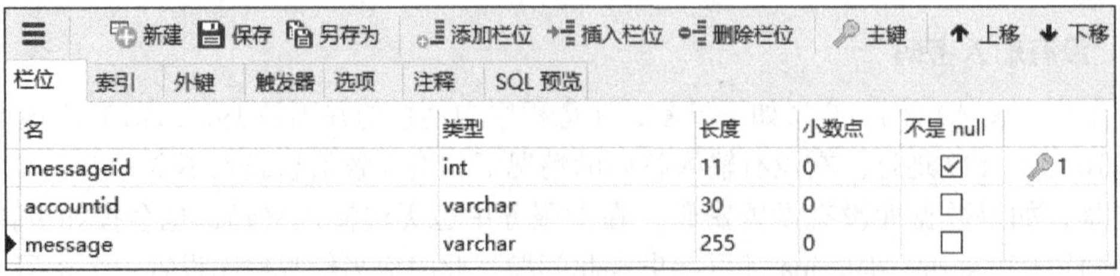

名	类型	长度	小数点	不是 null	
messageid	int	11	0	☑	🔑1
accountid	varchar	30	0	☐	
message	varchar	255	0	☐	

图 7-3

```java
    @Select("select account_id,message from 'message','account' where account_id=#{accountid2}")
    @Results({
        @Result(property="accountid",column="account_id"),
        @Result(property="message",column="message"),
        @Result(property="message",column="message")
    })
    public List<MessageVO> queryMessageByAccountid(@Param("accountid2") String accountid);

public class MessageVO {
    private String message;
    private String accountid;
    //省略 setter、getter 方法…
}

    @RequestMapping("/queryMessage/{accountid}")
    public List<MessageVO> queryMessage(@PathVariable String accountid){
        return accountDao.queryMessageByAccountid(accountid);
    }
```

多表查询效果如图 7-4 所示。

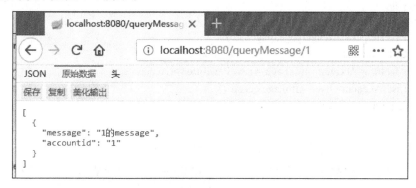

图 7-4

7.3.3 MyBatis 的注解式编程分页查询

MyBatis 的注解式编程分页查询通常有多种方式可以参考，如利用 PageHelper 插件或借助动态 SQL 进行分页。

1. PageHelper 分页查询

通过依赖 PageHelper 进行分页查询，pom.xml 文件增加依赖代码如下。

```xml
<!-- 分页插件 -->
<dependency>
    <groupId>com.github.pagehelper</groupId>
    <artifactId>pagehelper-spring-boot-starter</artifactId>
    <version>1.2.5</version>
</dependency>
```

在 application 资源配置文件中编写 PageHelper 的相关配置代码如下。

```yaml
pagehelper:
    helper-dialect: mysql
    reasonable: true
    support-methods-arguments: true
    params: count=countSql
    page-size-zero: true
```

7.2.4 节的 dao 层代码如下（原有 dao 层代码并无变换）。

```java
@Select("select * from 'account'")
@Results({
        @Result(id=true,property="accountid",column="account_id"),
        @Result(property="password",column="password"),
})
public List<Account> findAll();
```

在调用 findAll 前需用 PageHelper 进行封装，代码如下。

```java
public List<Account> selectAccounts(int page,int pageSize) {
    //page:第几页
    //pageSize:每页几条
        PageHelper.startPage(page, pageSize);
        List<Account> accounts = accountDao.findAll();
        PageInfo<Account> infoPage = new PageInfo<>(Accounts);
        List<Account> list = infoPage.getList();
        return list;
}
```

只查第 1 页第 1 条，分页查询后的效果如图 7-5 所示。

图 7-5

2．借助动态 SQL 进行分页

MyBatis 注解可以通过在动态 SQL 中传入相关分页信息达到分页效果，伪代码如下。

```java
@Select("select * from 'account' limit #{pageParam},#{pageSize}")
@Results({
    @Result(id=true,property="accountid",column="account_id"),
    @Result(property="password",column="password")
})
public List<Account> findAll(@Param("pageParam") String page,@Param("pageSizeParam") String pageSize);
```

只需要将 page 和 pageSize 作为入参参数或入参对象，传入 SQL 语句中即可。也可自行编写 MyBatisPage.java 对象，内部需含有分页、排序等相关参数，待频繁分页时，直接传入自定义的分页对象即可。

7.3.4 注册 DataSource 数据源

创建 Druid 数据源配置 DataSourcesConfig.java 代码如下。DataSourcesConfig.java 只需要编写一个带有@Configuration 注解的配置类即可，用@Bean 注解对 DataSource 进行注册，Spring 容器会以@Bean 注册的 DataSource 为主。本实例使用了阿里巴巴公司的 Druid 作为数据源进行使用。

```java
package com.zfx.config;
import javax.sql.DataSource;
import org.springframework.context.annotation.Bean;
import org.springframework.context.annotation.Configuration;
import com.alibaba.druid.pool.DruidDataSource;
@Configuration
public class DataSourcesConfig {
    @Bean
    public DataSource getDataSources() {
        DruidDataSource druidDataSource = new DruidDataSource();
        druidDataSource.setUrl("jdbc:mysql://192.168.102.128:3306/justLearn");
        druidDataSource.setDriverClassName("com.mysql.jdbc.Driver");
        druidDataSource.setUsername("root");
        druidDataSource.setPassword("root");
        return druidDataSource;
    }
}
```

@Configuration 注解修饰的函数可以使 Spring 容器对其进行扫描，@Bean 注解修饰的函数可以使 Spring 容器对 DataSource 进行注册，即直接省略了 application 资源配置文件中的相关内容，在 7.2 节中，直接删除 application 资源配置文件也可执行项目，运行效果如图 7-6 所示。

```
#spring.datasource.url=jdbc:mysql://192.168.102.128:3306/justLearn
#spring.datasource.username=root
#spring.datasource.password=
#spring.datasource.driver-class-name=com.mysql.jdbc.Driver
```

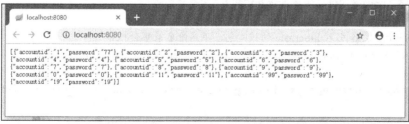

图 7-6

阿里巴巴公司的 Druid 连接池含有 Web UI 用来监控当前数据源的运行状态，省略其监控的部署方式，具体内容可参阅相关文档。

7.4 【实例】Spring Boot 整合 spring-data-jpa

7.4.1 实例背景

Spring Boot 整合 spring-data-jpa 通常使用两种方式，分别为 JPA 命名式和 JPA 注解式。JPA 注解式编程与 MyBatis 注解式编程相似。因为篇幅有限，且 JPA 命名式编程几乎不会出现在实际生产环境中，所以本实例将创建 boot_06_6 工程，只以 Spring Boot 使用 JPA 注解式方式整合 spring-data-jpa 实现增删改查等操作。

7.4.2 添加 Pom 文件

JPA 是 spring-data 的子项目之一，所以在 spring-boot-starter 中含有该项目，直接引入即可，不需要输入相关 version 信息，pom.xml 文件增加相关依赖代码如下。

```xml
<dependency>
    <groupId>org.springframework.boot</groupId>
    <artifactId>spring-boot-starter-data-jpa</artifactId>
</dependency>
```

7.4.3 添加资源配置文件中的相关信息

无论用哪个 ORM 框架整合数据库，Spring Boot 都需要使用数据库的密码，若数据库没有设置密码即无法书写密码，依旧会报 7.3.8 节中的错误。application.properties 资源配置文件代码如下。

```
spring.datasource.url=jdbc:mysql://192.168.138.141:3306/justLearn
spring.datasource.username=zhangfangxing
spring.datasource.password=zhangfangxing
spring.datasource.driver-class-name=com.mysql.jdbc.Driver
#指定数据库类型
spring.jpa.database=mysql
#控制台打印 sql
spring.jpa.show-sql=true
#建表策略，update 即根据实体更新表结构
spring.jpa.hibernate.ddl-auto=update
#表中字段命名策略，引入 hibernate 的核心包，不然命名策略会报错
spring.jpa.hibernate.naming.implicit-strategy=org.hibernate.boot.model.naming.ImplicitNamingStrategyLegacyJpaImpl
spring.jpa.hibernate.naming.physical-strategy=org.hibernate.boot.model.naming.PhysicalNamingStrategyStandardImpl
spring.jpa.hibernate.naming.strategy=org.hibernate.cfg.ImprovedNamingStrategy
#方言
```

```
spring.jpa.properties.hibernate.hbm2ddl.auto=validate
spring.jpa.properties.hibernate.dialect=org.hibernate.dialect.MySQL5InnoDBDialect
spring.jpa.hibernate.use-new-id-generator-mappings=false
```

7.4.4 添加实体类映射

为 spring-data-jpa 增加 Java Bean 实体类映射 User.java 代码如下。

```java
package com.zfx.bean;
import javax.persistence.Column;
import javax.persistence.Entity;
import javax.persistence.GeneratedValue;
import javax.persistence.GenerationType;
import javax.persistence.Id;
import javax.persistence.Table;
@Entity
@Table(name="user")
public class User {
    /** 用户 ID */
    @Id//主键
    @GeneratedValue(strategy = GenerationType.AUTO)//自动生成
    @Column(name="id",nullable = true)
    private int id;
    /** 用户编号 */
    @Column(name="user_no",nullable = true)
    private String userNo;
    /** 用户名称 */
    @Column(name="user_name",nullable = true)
    private String userName;
    /** 用户密码 */
    @Column(name="user_pwd",nullable = true)
    private String userPwd;
}
```

JPA 注解式编程与 MyBatis 注解式编程的区别是 MyBatis 不需要通过实体类映射数据库，因为 MyBatis 底层在通过打开 sqlSessionFactory 工厂得到数据库的驱动后，直接使用数据库驱动执行 SQL 语句，伪代码如下。

```java
public    void test(){
    InputStream inputStream = Resources.getResourceAsStream("sqlMapConfig.xml");
    SqlSessionFactory sqlSessionFactory = new SqlSessionFactoryBuilder().build(inputStream);
    SqlSession session = sqlSessionFactory.openSession();
    User user = session.selectOne("test.selectAccountById", 1);
}
```

而 JPA 底层是通过 Hibernate 实现的，执行 SQL 语句时 Hibernate 底层会根据反射找到 Java Bean 的全路径名，并通过 Java Bean 中的注解将 HQL 语句映射成 SQL 语句，从而实现对数据库的增删改查操作。

JPA 的 Java Bean 编写只需要使用@Entity 注解说明该类为实体类，通过@Table 注解说明映射的数据库表名，利用@Column 映射列名即可。

7.4.5 添加 JPA 的 dao 层

在 7.2.4 节中，Mybatis 注解式编程通过@Mapper 注解修饰 MyBatis 的 dao 层，原因是 MyBatis 需要将 dao 层的动态 SQL 转化成 MyBatis 的 XML 形式。但是 JPA 不需要将动态 SQL 转化成 XML 的过程，所以在 JPA 的 dao 层编写上，不需要任何注解修饰 JPA 的 dao 层类。JPA 的 UserDao.java 代码如下：

```java
package com.zfx.dao;
import java.util.List;
import org.springframework.data.jpa.repository.JpaRepository;
import org.springframework.data.jpa.repository.Modifying;
import org.springframework.data.jpa.repository.Query;
import org.springframework.data.repository.query.Param;
import org.springframework.transaction.annotation.Transactional;
import com.zfx.bean.User;
public interface UserDao extends JpaRepository<User, Integer>{

    @Query(value="select * from user",nativeQuery = true)
    public List<User> selectMyUsers();
    @Query(value="select * from user where user.id =:#{#userParam.id}",nativeQuery = true)
    public List<User> selectOneUserById(@Param("userParam") User user);
    @Query(value="from user u where u.id =?1",nativeQuery = true)
    public List<User> selectOneUserById2(int id);
    @Transactional
    @Modifying
    @Query(value="update user set user_name=:#{#userParam.userName} , user_pwd=:#{#userParam.userPwd} where id=:#{#userParam.id}",nativeQuery = true)
    public int updateOneUserById(@Param("userParam") User user);
    @Transactional
    @Modifying
    @Query(value="DELETE FROM user where id =:id",nativeQuery = true)
    public int deleteOneUserById(@Param("id")int id);
}
```

（1）@Transactional 注解为事务注解，后续会详细介绍。

（2）继承 JpaRepository<T,ID>工厂时，T 处编写相应的实体类名称，ID 处编写实体类的主键名称。

（3）动态 HQL 中的 "select *" 代码可以省略，如 selectOneUserById2 函数。

（4）动态 HQL 中的 "？1" 代码代表第一个入参。

（5）动态 HQL 中的 ":" 代表赋值，可以赋予@Param 注解修饰的值，":"后需使用@Param 注解的 name 名称，而不是入参的名称。

（6）动态 HQL 中的 delete 与 update 在 MyBatis 的编程中使用@Update、@Delete 注解，

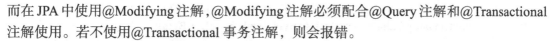

而在 JPA 中使用@Modifying 注解，@Modifying 注解必须配合@Query 注解和@Transactional 注解使用。若不使用@Transactional 事务注解，则会报错。

（7）动态 HQL 中不区分大小写。

（8）当 Mybatis 注解式编程入参为对象时，使用"#{EntityName.columnName}"进行入参，而在 JPA 中需使用":#{#EntityName.columnName}"进行入参，如 selectOneById 函数。

（9）JPA 需要通过其 JpaRepository 工厂类对 HQL 进行映射，因此需要继承 JPA 下任意一款 Repository 工厂。JpaRepository 工厂类写法不仅在 spring-data-jpa 中使用，spring-data-redis、spring-data-mongo 与 spring-data-jpa 的写法几乎相同。

（10）nativeQuery 选择 true 表示使用原 SQL 语句，选择 false 表示使用 HQL 语句。

7.4.6 添加 Controller 控制层查询 JPA 的 dao 层

为方便测试，创建 UserController.java 类，编写 index()函数返回 JPA 所查询的相关信息，其代码如下。

```java
package com.zfx.controller;
import java.util.List;
import org.springframework.beans.factory.annotation.Autowired;
import org.springframework.web.bind.annotation.RequestMapping;
import org.springframework.web.bind.annotation.RestController;
import com.zfx.bean.User;
import com.zfx.dao.UserDao;
@RestController
public class UserController {
    @Autowired
    private UserDao userDao;

    @RequestMapping({"/","/index"})
    public Object index(){
        List<User> selectMyUsers = userDao.selectMyUsers();
        System.out.println("所有的用户信息： "+selectMyUsers);

        User user = new User();
        user.setId(2);
        List<User> selectOneUserById = userDao.selectOneUserById(user);
        System.out.println("得到 ID 为"+2+"的用户信息是： "+selectOneUserById);

        List<User> selectOneUserById2 = userDao.selectOneUserById2(2);
        System.out.println("得到 ID 为"+2+"的用户信息是： "+selectOneUserById2);

        User user2 = new User();
        user2.setId(2);
        user2.setUserName("myUsername222");
        user2.setUserPwd("myPwd222");
```

```
        int updateOneUserById = userDao.updateOneUserById(user2);
        System.out.println("updateOneUserById"+updateOneUserById);

        int deleteOneUserById = userDao.deleteOneUserById(2);
        System.out.println("deleteOneUserById"+deleteOneUserById);

        return selectMyUsers;
    }
}
```

因为 JPA 的工厂会自动管理 JPA 相关内容，并存入 Spring 容器中，所以不需要在 JPA 的 dao 层编写 @Service 等注解，可直接使用@Autowried 注解将 JPA 的 dao 层进行注入 Controller 中。

7.4.7 当前项目结构

JPA 的操作原理十分简单，只需要编写 User、UserDao、application 三个文件即可，当前项目结构如图 7-7 所示。

7.4.8 运行结果

相关数据执行正常，不再赘述。但 JPA 不具备该数据原本不存在于数据库就新增的能力，此时 JPA 会返回 0，即没有任何修改的行。而 MyBatis 的@Update 注解具备上述能力。

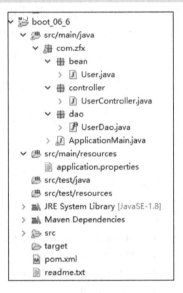

图 7-7

```
Hibernate: select * from user
所有的用户信息： [User [id=123, userNo=123, userName=123, userPwd=123]]
Hibernate: select * from user where user.id =?
得到 ID 为 2 的用户信息是： []
Hibernate: select * from user u where u.id =?
得到 ID 为 2 的用户信息是： []
Hibernate: update user set user_name=? , user_pwd=? where id=?
updateOneUserById0
Hibernate: DELETE FROM user where id =?
deleteOneUserById0
```

7.4.9 实例易错点

1．修改时没有使用事务注解

spring-data-jpa 的修改和删除必须配合@Transactional 事务注解使用，否则会报下述错误，Mybatis 注解式编程的修改和删除则不需要。

```
javax.persistence.TransactionRequiredException: Executing an update/delete query
```

2. 入参对象的符号错误或未写入参

EL1008E 和 EL1007E 是两种 dao 层函数中未写入参或入参符号错误可能出现的报错内容。SQL 类错误的原因较多，下面只列举两种。

> EL1008E: Property or field 'userParam' cannot be found on object of type 'java.lang.Object[]' - maybe not public?
>
> EL1007E: Property or field 'userName' cannot be found on null

3. 表或列映射错误

JPA 无法创建 entityManagerFactory 实体类工厂时，大多因为有部分表或列的映射错误。

> Error creating bean with name 'entityManagerFactory' defined in class path resource [org/springframework/boot/autoconfigure/orm/jpa/HibernateJpaConfiguration.class]: Invocation of init method failed; nested exception is javax.persistence.PersistenceException: [PersistenceUnit: default] Unable to build Hibernate SessionFactory
> ###

@table 注解中的表名和数据库的表名映射不上，导致无法正常创建实体类工厂。从上文错误向下追查可发现：

> Schema-validation: missing table [User]

@Column 注解中的列名和数据库的列名映射不上，导致无法正常创建实体类工厂。从上文错误向下追查可发现：

> Schema-validation: missing column [id2] in table [user]

7.5 本章小结

本章通过 Spring Boot 微服务整合 MyBatis 注解和 JPA 注解，达到操作数据库进行增删改查操作的效果。

由于 JPA 注解、MyBatis 注解的分页和多表十分类似，本章不再重复介绍。通过运用 7.2 节和 7.4 节中的@Select、@Query 等注解，可了解利用注解进行数据库开发的思想。

7.6 习题

1. 创建 Spring Boot 工程，使用 MyBatis 注解整合 MySQL 数据库，执行增删改查等相关命令。

2. 创建 Spring Boot 工程，使用 spring-data-jpa 整合 MySQL 数据库，分别使用 HQL 和 SQL 执行增删改查等相关命令。

第 8 章 微服务事务

当前事务中所有增删改查都执行成功后,事务会进行提交,所改动的数据生效。当前事务中所有增删改查的数据任意一条没有正常修改,当前事务会进行回滚,其他修改过的数据都不再生效,将会改成修改前的数据。

微服务事务主要依靠 7.4.9 节中提到的@Transactional 注解,本章将详细介绍。

8.1 @Transactional 注解

对于编程式事务管理,Spring 使用 TransactionTemplate 注解进行事务管理。

对于声明式事务管理,Spring 使用@Transactional 注解进行事务管理。

@Transactional 注解源自 org.springframework.transaction.annotation 包,原 SSM 架构需要使用@EnableTransactionManagement 注解,开启对声明式事务@Transactional 注解的支持。由于 Spring Boot 默认开启声明式事务,所以不需要编写@EnableTransactionManagement 注解即可直接使用@Transactional 注解。

@Transactional 注解可以编写在函数上,也可以编写在类上,若@Transactional 注解对类进行修饰的话,则说明该类下所有用 public 修饰的函数都使用了声明式事务。

@Transactional 注解配置的参数及释义如表 8-1 所示。

表 8-1 @Transactional 注解配置的参数及释义

参数	释义	参数默认值
Value	transactionManager 参数的别名,释义与其相同	""
transactionManager	事务管理器的名称,即 Spring Bean 名称	""
propagation	事务传播行为	Propagation.REQUIRED
isolation	事务隔离级别	Isolation.DEFAULT
timeout	事务超时时间	TransactionDefinition.TIMEOUT_DEFAULT
readOnly	是否只读	false
rollbackFor	指定 0 个到多个异常传播行为必须进行事务回滚	{}
rollbackForClassName	指定 0 个到多个异常名称必须进行事务回滚	{}
noRollbackFor	指定 0 个到多个异常传播行为绝不进行事务回滚	{}
noRollbackForClassName	指定 0 个到多个异常名称绝不进行事务回滚	{}

8.1.1 @Transactional 声明式事务的传播行为

如表 8-1 所示，@Transactional 注解的事务传播行为依靠 propagation 参数类进行配置，propagation 参数通常以 org.springframework.transaction.annotation.Propagation 的枚举类进行配置，默认为 Propagation.REQUIRED。

事务传播行为决定了事务的执行管理方式，如下所述。

Propagation.REQUIRED 传播行为：支持当前事务，若不存在则创建新事务。例如，当 dao 层的 insert 函数被调用时，若没有事务则会新增一个事务；但在@Service 中，如果调用 insert 函数后又调用了 update 函数，在 update 函数没有事务的情况下，不会新增事务，update 函数则会沿用 insert 的事务，一旦发生任何报错，就全部进行回滚。

Propagation.SUPPORTS 传播行为：支持当前事务，若不存在则不新建事务，按照非事务性的增删改执行与操作。例如，当 dao 层的 insert 函数被调用时，若没有事务则不会新增事务，直接进行操作，在数据库层直接进行 commit 提交，发生任何情况都不会回滚数据。

Propagation.MANDATORY 传播行为：支持当前事务，若不存在则抛出异常并报错。例如，当 dao 层的 insert 函数被调用时，若没有事务，则应用程序会直接抛出异常，不会在数据库中进行增删改操作。

Propagation.REQUIRES_NEW 传播行为：创建新事务，若当前存在事务，则挂起该事务。例如，当 dao 层的 insert 函数和 update 函数被调用时分别创建两个事务，若 update 函数中的事务出现了问题，只回滚 update 函数中的事务，而不会回滚 insert 函数中的事务。

Propagation.NOT_SUPPORTED 传播行为：不用事务执行操作，若当前存在事务，则挂起该事务。例如，当 dao 层的 insert 函数被调用时，insert 函数含有事务，但是被调用时强制不会进行任何事务管理，将事务暂时挂起，直到 insert 提交结束修改后，才会释放事务。

Propagation.NEVER 传播行为：不用事务执行操作，若存在事务，则抛出异常并报错。例如，当 dao 层的 insert 函数被调用时，若有事务，则应用程序会直接抛出异常，不会在数据库中进行增删改操作。

Propagation.NESTED 传播行为：若当前含有事务，则创建一个事务作为当前事务的嵌套事务来运行；若当前没有事务，则该取值等价于 PROPAGATION_REQUIRED。注意，此传播行为不支持 JPA 和 Hibernate。

8.1.2 脏读、不可重复读与幻读

1. 脏读（Dirty Read）

脏读针对未提交事务的数据。如果事务 1 正在修改一行数据，而事务 2 开始读取该行数据，由于事务 1 没有修改结束，事务 2 已经读取了数据并返回，且传输给了用户，事务 2 返回的数据则是未修改完成的数据，该情景被称为脏读。在脏读下读取到的数据有可能是失效数据。例如，判断某用户是否有某权限，需读取某字段为"Y"或"N"，脏读得到的"N"

可能是正在被修改且还没修改完成的，可前台获得数据依旧无此权限。在高并发情况下的商品数量也经常会有脏读的情况产生。

2．不可重复读（NonRepeatable Read）

不可重复读针对提交事务前后数据的对比。例如，事务 1 中含有以下步骤：第一步读取了某行，第二步修改了某行，第三步重新读取了某行，该事务较常见，但是由于同一个事务中第一步读取和第三步读取的同一行数据并不相同，就产生了不可重复读情况。在不同事务的情况下，也容易出现不可重复读情况。

3．幻读（Phantom Read）

幻读针对提交事务前后数据数目的对比。例如，事务 1 中含有以下步骤：第一步读取了某表的一串结果集，第二步增加或删除了该表的数据，第三步重新读取了某表的一串结果集，但是由于同一个事务中第一步读取和第三步读取的结果集的数目并不相同，就产生了幻读情况。

8.1.3 @Transactional 声明式事务的隔离级别

如表 8-1 所示，@Transactional 注解的事务隔离级别依靠 isolation 参数类进行配置，isolation 参数通常以 org.springframework.transaction.annotation.Isolation 的枚举类进行配置，默认为 Isolation.DEFAULT。

事务的隔离级别决定了事务的完整性，解决了脏读、不可重复读、幻读等并发问题，如下所述。

Isolation.DEFAULT 隔离级别：使用数据库中默认的隔离级别。

Isolation.READ_UNCOMMITTEDT 隔离级别：允许读取一个由事务修改的行。在修改该行的过程中，可以读取修改后的数据。易引起脏读、不可重复读、幻读等情况。

Isolation.READ_COMMITTEDTED 隔离级别：与 Isolation.READ_UNCOMMITTEDT 隔离级别相反，即不允许读取任何一个正在进行事务的行。必须等待该行事务提交后才能读取。可以避免出现脏读，但可能会出现不可重复读和幻读。

Isolation.REPEATABLE_READ 隔离级别：禁止事务读取其中有未提交更改的行。在 Isolation.READ_COMMITTEDTED 隔离级别中，如果事务 1 修改了一行，在事务 1 没有提交前，事务 2 不允许读取该行。在 Isolation.REPEATABLE_READ 隔离级别中，如果事务 1 读取了一条事务，在事务 1 没有返回结束的情况下，事务 2 不允许更改该行。可以避免出现脏读和不可重复读，但可能出现幻读。

Isolation.SERIALIZABLE 隔离级别：采用顺序执行的方式。如果事务 1 修改了一行，事务 2 会等待事务 1 提交后再读取。如果事务 1 读取了一行，事务 2 会等待事务 1 正确返回后再修改。避免了 Isolation.READ_COMMITTEDTED 隔离级别和 Isolation.REPEATABLE_READ 隔离级别的缺陷。避免出现脏读、不可重复读、幻读。但由等待次数、时间等原因，该隔离级别的开销过大，只在极少情况，不在乎任何系统性能，绝对强调系统的强一致性时才使用。

由表 8-2 可更直观地看出@Transactional 声明式事务的隔离级别与脏读、不可重复读、幻读之间的关系。

表 8-2　@Transactional 声明式事务的隔离级别与脏读、不可重复读、幻读之间的关系

隔 离 级 别	脏读	不可重复读	幻读
Isolation.READ_UNCOMMITTEDT	可能	可能	可能
Isolation.READ_COMMITTEDTED	不可能	可能	可能
Isolation.REPEATABLE_READ	不可能	不可能	可能
Isolation.SERIALIZABLE	不可能	不可能	不可能

级别越高，性能越差，隔离性越高。

8.1.4　@Transactional 声明式事务的超时时间

如表 8-1 所示，@Transactional 注解的事务隔离级别依靠 timeout 参数类进行配置，默认为 TransactionDefinition.TIMEOUT_DEFAULT，该常量为−1，永不超时。自定义该参数后，若增删改时间超过自定义参数，则直接将事务进行回滚，不会提交事务。

8.1.5　@Transactional 声明式事务的只读

如表 8-1 所示，@Transactional 注解的事务隔离级别依靠 readOnly 参数类进行配置，readOnly 参数的配置如下。

readOnly=true 表明所注解的方法或类只读取数据。

readOnly=false 表明所注解的方法或类增加、删除、修改数据。

伪代码如下。

```
@Transactional(readOnly = true)
public class DefaultFooService implements FooService {
public Foo getFoo(String fooName) {
    // do something
}
}
```

若采用 readOnly=true 的方式，则不允许对数据进行增删改操作；若只读取，则通过 Hibernate 的二级缓存读取相关数据。例如，某功能通过 readOnly 读取数据，但在读取的过程中其他数据修改并提交了事务，readOnly 再次读取时有可能读取 Hibernate 的二级缓存，即旧数据。但在只读取的情况下，readOnly=true 的性能优于 readOnly=false。

如果用 readOnly=true 修饰 update 函数，伪代码如下。

```
@Transactional(readOnly=true)
@Modifying
@Query(value="update user set user_name=:#{#userParam.userName} , user_pwd=:#{#userParam.userPwd} where id=:#{#userParam.id}",nativeQuery = true)
public int updateOneUserById(@Param("userParam") User user);
```

强行调用该函数，后台报错如下。

java.sql.SQLException: Connection is read-only. Queries leading to data modification are not allowed

强行调用该函数，前台报错如图 8-1 所示，此处提示报错且无法进行增删改操作。

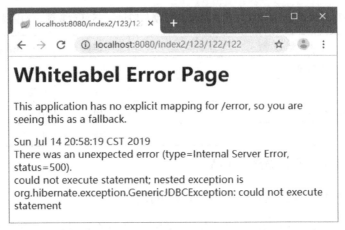

图 8-1

8.1.6　@Transactional 声明式事务指定异常

如表 8-1 所示，rollbackFor、rollbackForClassName、noRollbackFor、noRollbackForClassName 四个参数十分类似，在此统一讲解。

rollbackFor 指定回滚异常参数与 noRollbackFor 指定不回滚异常参数，都需要使用异常.class 反射；而 rollbackForClassName 指定回滚异常参数与 noRollbackForClassName 指定不回滚异常参数，都需要使用异常的名称，伪代码如下。

```
@Transactional(readOnly=false,
        rollbackFor=SQLException.class,
        rollbackForClassName="SQLException",
        noRollbackFor=RuntimeException.class,
        noRollbackForClassName="RuntimeException")
@Modifying
@Query(value="update user set user_name=:#{#userParam.userName}，user_pwd=:#{#userParam.userPwd} where id=:#{#userParam.id}",nativeQuery = true)
public int updateOneUserById(@Param("userParam") User user);
```

使用 noRollbackFor 或 rollbackForClassName 后（其一即可，上述代码只展示效果），一旦遇到 RuntimeException 报错，整个事务依旧不会回滚。默认情况下发生任何报错都会回滚。

8.2　【实例】Spring Boot 整合声明式事务

8.2.1　实例背景

本实例将修改 7.4 节的 boot_06_6 工程，使用@Transactional 注解对其进行事务管理，若在 service 中发生错误,则让事务直接回滚。User.java、UserDao.java、ApplicationMain.java 等类与 7.4 节中编写的相关类相同,此处简略。

8.2.2 整合 @Transactional 的 Service 层编写

整合 @Transactional 的 Service 层代码如下,UserServiceImpl 提供了两个函数,在 service1 函数中发生任何错误都会回滚;在 service2 函数中,只有运算异常不回滚,其他异常同样回滚。

```java
package com.zfx.service.impl;
import org.springframework.beans.factory.annotation.Autowired;
import org.springframework.stereotype.Service;
import org.springframework.transaction.annotation.Transactional;
import com.zfx.bean.User;
import com.zfx.dao.UserDao;
import com.zfx.service.UserService;
@Service
public class UserServiceImpl implements UserService{

    @Autowired
    private UserDao userDao;

    @Transactional(rollbackFor=Exception.class)
    public int service1(User user2){
        int updateOneUserById = userDao.updateOneUserById(user2);
        int i = 1/0;
        System.out.println("updateOneUserById"+updateOneUserById);
        return updateOneUserById;
    }
    @Transactional(noRollbackFor=java.lang.ArithmeticException.class)
    public int service2(User user2){
        int updateOneUserById = userDao.updateOneUserById(user2);
        int i = 1/0;
        System.out.println("updateOneUserById"+updateOneUserById);
        return updateOneUserById;
    }
}
```

在 service1 和 service2 的 dao 层函数中,由于 Service 层的函数用 @Transactional 注解修饰函数,所以即使 dao 层中的函数不再使用 @Transactional 修饰 dao 层函数,JPA 也不会报 7.4.9 节中的错误,dao 层代码如下。

```java
@Modifying
@Query(value="update user set user_name=:#{#userParam.userName} , user_pwd=:#{#userParam.userPwd} where id=:#{#userParam.id}",nativeQuery = true)
public int updateOneUserById(@Param("userParam") User user);
```

```
@Modifying
@Query(value="DELETE FROM user where id =:id",nativeQuery = true)
public int deleteOneUserById(@Param("id")int id);
```

如果 Service 层和 dao 层函数使用了@Transactional 注解，将会采用事务嵌套的方式。

嵌套事务相当于函数 A 调用函数 B，且函数 A 与函数 B 都被事务修饰，此时嵌套事务不再关心外层事务的传播行为，而只关心内部事务的传播行为。嵌套事务的逻辑如表 8-3 所示。

表 8-3 嵌套事务的逻辑

内层事务传播行为	嵌套事务释义	结　　论
Propagation.REQUIRED	内外层方法共用外层方法的事务	有异常一起回滚，无异常一起提交
Propagation.REQUIRES_NEW	当执行内层被调用方法时，外层方法的事务会挂起；两个事务相互独立，不会相互影响	外层调用方法出现异常后，会回滚，但不会回滚嵌套事务内层被调用方法

8.2.3 整合@Transactional 的 Controller 层编写

为方便测试，创建 IndexController.java 接口类，其部分代码如下。Controller 层只负责调用 Service，service1 和 service2 中都出现了 ArithmeticException 异常，但由于 service1 函数的事务中出现任何异常都会回滚，而 service2 函数的事务中出现 ArithmeticException 异常不会回滚。因此 controller1 函数无法修改数据库中的信息，而 controller2 函数即使内部 service2 报错，service2 函数也不会回滚。

```
@RequestMapping({"/service1/{id}/{username}/{password}"})
public Object controller1(
        @PathVariable("username") String username,
        @PathVariable("password") String password,
        @PathVariable("id") Integer id) throws Exception{

    User user2 = new User();
    user2.setId(id);
    user2.setUserName(username);
    user2.setUserPwd(password);
    int service1 = 0;
    try {
        service1 = service.service1(user2);
    } catch (Exception e) {
        e.printStackTrace();
    }
    return service1;
}
@RequestMapping({"/service2/{id}/{username}/{password}"})
public Object controller2(
        @PathVariable("username") String username,
        @PathVariable("password") String password,
```

```
        @PathVariable("id") Integer id) throws Exception{

    User user2 = new User();
    user2.setId(id);
    user2.setUserName(username);
    user2.setUserPwd(password);
    int service1 = 0;
    try {
        service1 = service.service2(user2);
    } catch (Exception e) {
        e.printStackTrace();
    }
    return service1;
}
```

8.2.4 当前项目结构

当前项目结构如图 8-2 所示，与 7.4 节中的结构基本相同。

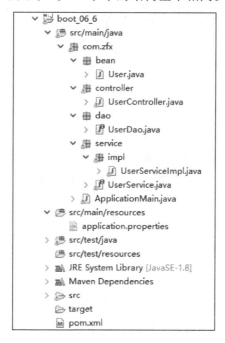

图 8-2

8.2.5 运行结果

直接运行 IndexController.java 中的接口 localhost:8080/service1/123/133/133，观察返回结果。如图 8-3 所示，使用 service1 接口并未更改数据库中的相关数据。

如图 8-4 所示，使用 service2 接口返回了 0，代表未修改任何行，但数据库中依旧修改了相关数据。

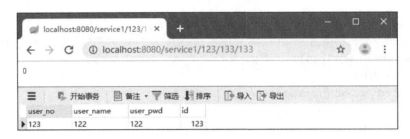

图 8-3

图 8-4

在图 8-3 与图 8-4 的操作后，后台日志均如下。

Hibernate: update user set user_name=? , user_pwd=? where id=?
java.lang.ArithmeticException: / by zero

8.2.6 实例易错点

1．数据库事务与 Spring 事务的关系

Spring 事务是以数据库事务作为基础设计的。若如果数据库引擎不支持事务，则在应用程序中编写任何事务都将无效。例如，MySQL 的 InnoDB 数据库引擎支持事务，MyIsam 数据库引擎不支持事务。

2．@Transactional 注解修饰在类头上

@Transactional 注解只修饰在类头上时，只对 public 修饰的函数生效，若其中有非 public 修饰的函数，则@Transactional 不会对其进行管理。

3．@Transactional 注解修饰的函数所在位置

@Transactional 注解修饰的函数不能与调用函数处于同一个类中，因为@Transactional 注解是声明，事务的管理器实现类依靠 AOP 扫描声明后，使用代理类对函数进行统一管理，@Transactional 注解的设计方式与@Async 注解、@CacheResult 注解基本相同。可参考 5.4.4 节和 6.2 节。

4．tryCatch 不可异常

如果 tryCatch 异常，应用程序将认为该异常已被捕捉，捕捉即处理。因此@Transactional 认为已处理的异常不是错误，不会对已处理的异常进行回滚。如下代码是不会回滚的代码。

```
@Transactional(rollbackFor=Exception.class)
public int service1(User user2){
    int updateOneUserById = userDao.updateOneUserById(user2);
    try {
        int i = 1/0;
    } catch (Exception e) {}
    System.out.println("updateOneUserById"+updateOneUserById);
    return updateOneUserById;
}
```

8.2.3 节中的代码，将 tryCatch 放置在 Controller 中，service 中出现了异常既没有捕捉也没有抛出，所以@Transactional 会对其进行管理。

5. 部分异常未回滚

@Transactional 注解默认回滚 Error 类异常，当应用程序出现部分非 Error 的警告类异常，或 RuntimeException()、SQLtimeException()超时类异常时，@Transactional 注解不会对其回滚，解决方式如下。

```
@Transactional(rollbackFor=Exception.class)
```

8.3 本章小结

通过 8.2 节的实例更加完整地了解@Transational 注解，包括该注解在@Service 层对多个 dao 层的应用方式，扩展了第 7 章的内容，希望通过本章可以使读者更加清晰地理解事务的使用方式、传播行为、隔离级别等内容。

8.4 习　　题

1. 创建 Spring Boot 工程，使用 spring-data-jpa 整合 MySQL 数据库，执行增删改查等相关命令。

（1）区分使用 Innodb 引擎和 Myisam 引擎是否相同。

（2）执行途中强制报错，记录结果。

（3）执行途中关闭服务器，记录结果。

（4）执行途中关闭数据库，记录结果。

（5）使用 Swagger 执行增删改查途中强制报错，结果与正常接口是否相同。

2. 创建 Spring Boot 工程 1、Spring Boot 工程 2、Spring Boot 工程 3 与 Spring Boot 工程 3-2，使用 Spring Boot 工程 1 调用 Spring Boot 工程 2，Spring Boot 工程 2 内部使用 @Async+Ribbon 调用 Spring Boot 工程 3 或 Spring Boot 工程 3-2，此时 Spring Boot 工程 2 执行报错，程序有何反应。

（1）区分被 trycatch 和非 trycatch 两种情况是否相同。

（2）区分有@Async 异步线程池和无@Async 异步线程池两种情况是否相同。

（3）若报错点为 Spring Boot 工程 1、Spring Boot 工程 3 或 Spring Boot 工程 3-3，结果是否相同。

第 9 章　微服务的缓存与分布式的消息通信

第 7 章通过 Spring Boot 整合 MySQL 达到增删改查关系型数据库的目的，通过第 8 章对增删改查进行事务管理。但是关系型数据库在被大量并发调用时，性能堪忧。本章通过 Redis 缓存与分布式通信解决高并发场景和大量数据的预读能力问题，总体提高了微服务的性能。

9.1　Redis

Redis 是一个开源的（BSD 许可的）内存数据结构存储，作为数据库、缓存和消息代理的非关系型数据库，支持字符串、散列、列表、集合、带范围查询的排序集、位图、超日志、具有 RADIUS 查询和流的地理空间索引等数据结构。Redis 内置了复制、Lua 脚本、事务处理和不同级别的磁盘上持久化，通过 Redis Sentinel 和 Redis 集群自动分区提供了高可用性。Lua 是一种小巧的语言，作为脚本语言而存在，通常在 Unity3d、Cocos2d、服务器运维等方面独立存在着。

Redis 使用类似于 Map 的 Key-Value 数据类型进行存储。与其类似的项目还有 Memcached、MongoDB。为了保证效率，数据都是存储在内存中的。与实际内存的区别是，Redis 会周期性地把更新的数据写入磁盘，或修改操作写入追加的记录文件，并以此为基础实现了 Master-Slave 主从同步的模式，后续又有优化，可部署为哨兵模式。Redis 提供了将缓存持久化的能力，在不同的场合下可以对关系型数据库起很好的补充作用。Redis 使用广泛，除 Java 外，它提供了 C、C++、C#、PHP、Perl、Object-C、Python、Ruby 等语言的整合方式，使用起来十分方便，并且也是最流行的非关系型数据库之一。Spring Boot 通常整合 Redis、Mongo DB 等非关系型数据库作为微服务的缓存。

9.1.1　BSD 协议

BSD 协议是一个给予使用者很大自由的开源协议。以 BSD 协议代码为基础进行二次开发自己的产品，需要满足以下 3 个条件。

（1）若再发布的产品包含源代码，则在源代码中必须带有原来代码中的 BSD 协议。若再发布的只是二进制类库/软件，则需要在类库/软件的文档和版权声明中包含原来代码中的 BSD 协议。

（2）不可用开源代码的作者/机构名字和原来产品的名字进行市场推广。

（3）BSD 协议鼓励代码共享，但需要尊重代码作者的著作权。BSD 协议由于允许使用者修改和重新发布代码，也允许使用或在 BSD 协议代码上开发商业软件发布和销售，所以是对商业集成很友好的协议。大部分企业在选用开源产品时都首选 BSD 协议，因为可以完全控制第三方的代码，在必要时可以修改或二次开发。

9.1.2　Java 与 Redis 的历史

使用 Redis 最主要的目的是给关系型数据库提高存储与并发的能力。例如，要访问某张表时，可以先检查其是否存储于 Redis 中，若不存在，则访问关系型数据库 Oracle、MySQL 等，并且将读取到的数据重新存储到 Redis 中，下次访问可直接从 Redis 中读取。当然，在小项目中并不明显，但是若同时有几十万乃至上百万的访问量，其对性能的提升是飞跃的，极大减少了对数据库的压力。

注意，每当数据库中的数据改变时，必须改变 Redis 缓存中的数据，否则将会出现缓存与数据库不同步的现象，类似于脏读。因此使用 Redis 要注意关于分页查询时的解决方案，因为业务不同，所以可能使用的解决方案也不同。但大多可以采用 Spring AOP 的方式，从存入 Redis 缓存的 Key 名中进行分析，以取到相应结果。若 Java 应用程序的并发过高，也可因为业务，弱化 Redis 与 MySQL 间的强一致性（在高并发过程中可以 Redis 为主，并不存在 MySQL 中，或以生成对账文件的方式）只保证数据的最终一致性与最终同步，并保证对账时的完整性。

Redis 是 Java 项目中最常用的缓存之一，也是面试必备技能之一。Redis 有多种使用方式，也有 Spring Data Redis 和 Spring Cache 两种方式去整合 Redis，使 Redis 项目可以更加便捷地集成到项目中，而不需要使用手写 Connect 连接 Redis 数据库等方式。

9.1.3　Spring Data Redis

Spring 项目通过 Spring Data Redis 框架，可以将 Java Bean 数据使用键值对数据的格式存入 Redis 中。Spring Data Redis 提供了一个"模板"作为发送和接受消息的高级抽象化载体。Spring Data Redis 与 Spring 框架中的 JDBC 有很多相似之处。

Spring Data Redis 给 Spring 相关的应用程序提供了轻松的配置和访问能力，在使用过程中，用户不需要知道底层如何实现，只需要关注自身的业务即可。Spring Data Redis 依靠 Spring IOC 的方式，将 RedisTemplate 注入自身的应用程序中以操纵 Redis 数据库。

Spring Data Redis 使用 RedisConnection 为通信提供了核心的构建模块，以处理与 Redis 的连接通信，此时相当于将 Redis 连接中的 Connect 工具类进行了整合。若有特殊需要，RedisConnection 也提供了 getNativeConnection 返回通信的原始底层对象。在实际场景中，Spring Data Redis 提供 RedisConnectionFactory 的工厂类，统一管理 RedisConnection 连接，通常将 Redis 相关的配置数据存入 Spring Boot 的 application.properties 中，RedisConnection-Factory 可直接使用以上相关配置信息。整个过程十分便捷。其中最简单的使用方法是通过 IOC 容器将 RedisConnectionFactory 连接器工厂注入使用的类中。

9.2 【实例】微服务整合 Spring Data Redis 增删改查

9.2.1 实例背景

本实例创建 boot_07_1 工程,将 Spring Boot 项目使用 Spring Data Redis 整合 Redis 缓存,通过 RedisConnectionFactory 连接池工厂生成相应的 RedisTemplate 操作模板,依靠 RedisTemplate 操作模板对 Redis 进行增删改查的操作。创建项目后,在 pom.xml 文件中增加相应的 Spring Data Redis 依赖代码如下。

```xml
<!-- 整合 redis -->
<dependency>
    <groupId>org.springframework.boot</groupId>
    <artifactId>spring-boot-starter-data-redis</artifactId>
</dependency>
<dependency>
    <groupId>org.springframework.boot</groupId>
    <artifactId>spring-boot-starter-web</artifactId>
</dependency>
```

9.2.2 编写 application.properties 资源配置文件

编写 Spring Boot 使用 Spring Data Redis 整合 Redis 的 application.properties 资源配置文件代码如下。以下 Key 值命名都是 Spring Boot 已设计的,只要编辑好 Key 值,Spring Boot 读到此处数据就会自动处理与 RedisConnectionFactory 之间的联系,使 RedisConnectionFactory 连接 Redis 数据库,后续只要配置相应的 RedisTemplate 模板即可。

```properties
#redis 默认的数据库索引
spring.redis.database=0
#redis 的服务器地址
spring.redis.host=192.168.138.149
#redis 的服务器连接端口
spring.redis.port=6379
#redis 服务器连接密码(默认为空)
spring.redis.password=
#连接池最大连接数(使用复制表示没有限制)
spring.redis.pool.max-active=8
#连接池最大阻塞等待时间(使用复制表示没有限制)
spring.redis.pool.max-wait=-1
#连接池中最大的空闲连接
spring.redis.pool.max-idle=8
#连接池中最小的空闲连接
spring.redis.pool.min-idle=0
#连接超时时间(毫秒)
spring.redis.timeout=0
```

9.2.3 配置 RedisTemplate 模板

新建 Redis 的配置类 RedisCacheConfig.java 代码如下,在 RedisCacheConfig.java 配置类中给 RedisTemplate 模板注入 RedisConnectionFactory 连接工厂,通过 Jackson2JsonRedisSerializer 提供的序列化方式,将 Java Bean 对象通过 RedisTemplate 模板进行序列化并存入 Redis 中。

对象序列化机制允许将实现序列化的 Java 对象转换成字节序列,转换后的序列被保存到磁盘上,或通过网络进行传输。而在 Java 对象的序列化过程中,可以将 Java Bean 对象写入 IO 流中,使 Redis 从 IO 流中接收 Java Bean 对象。若需要使某个对象支持序列化机制,则必须使它的类是可序列化的,即该 Java Bean 需已实现 Serialize 接口。

Java 序列化的方式是将所有保存到磁盘的对象都有一个序列化的编号。当程序试图序列化一个对象时,程序会先检查该对象是否已经被序列化了,只有当该对象没有被序列化,系统才会将对象转换成字节流并输出;如果某个对象已经被序列化了,程序将只输出一个序列化编号,而非再次序列化该对象并对其进行输出。当对某个对象进行序列化时,系统会自动将该对象的所有属性依次进行序列化,如果某个属性是另一个对象,则另一个对象也需要实现 Serialize 接口,否则 Java 应用程序会报错,多个属性多次序列化的方式也称为递归序列化。

```
package com.zfx.config;

import org.springframework.context.annotation.Bean;
import org.springframework.context.annotation.Configuration;
import org.springframework.data.redis.connection.RedisConnectionFactory;
import org.springframework.data.redis.core.RedisTemplate;
import org.springframework.data.redis.serializer.Jackson2JsonRedisSerializer;
import org.springframework.data.redis.serializer.StringRedisSerializer;
import com.fasterxml.jackson.annotation.JsonAutoDetect;
import com.fasterxml.jackson.annotation.PropertyAccessor;
import com.fasterxml.jackson.databind.ObjectMapper;

@Configuration
public class RedisCacheConfig {
    /**
     * redisTemplate 序列化使用的 jdkSerializeable, 存储二进制字节码, 所以自定义序列化类
     * @param redisConnectionFactory
     * @return
     */
    @Bean("redisTemplate")
    public RedisTemplate<Object, Object> redisTemplate(RedisConnectionFactory redisConnectionFactory) {
        RedisTemplate<Object, Object> redisTemplate = new RedisTemplate<>();
        redisTemplate.setConnectionFactory(redisConnectionFactory);

        // 使用 Jackson2JsonRedisSerialize 替换默认序列化
```

```java
        Jackson2JsonRedisSerializer<Object> jackson2JsonRedisSerializer = new Jackson2JsonRedisSerializer<Object>(Object.class);

        ObjectMapper objectMapper = new ObjectMapper();
        objectMapper.setVisibility(PropertyAccessor.ALL, JsonAutoDetect.Visibility.ANY);
        objectMapper.enableDefaultTyping(ObjectMapper.DefaultTyping.NON_FINAL);

        jackson2JsonRedisSerializer.setObjectMapper(objectMapper);

        // 设置 Value 的序列化规则和 Key 的序列化规则
        redisTemplate.setValueSerializer(jackson2JsonRedisSerializer);
        redisTemplate.setKeySerializer(new StringRedisSerializer());
        redisTemplate.afterPropertiesSet();
        return redisTemplate;
    }
}
```

RedisTemplate 类：RedisTemplate 帮助 Java 应用程序简化 Redis 数据访问的代码操作，默认情况下，它使用对象的 Java 序列化是通过 jdkSerialtaseReDeSerialStase 实现的，对于大量字符串操作时可以考虑使用 StringRedisTemplate。RedisTemplate 一旦配置完成，RedisTemplate 的线程即是安全的。RedisTemplate 也是 Spring Data Redis 中最重要的一个类。未来增删改查都会由 RedisTemplate 进行操作。

Jackson2JsonRedisSerializer 类：用来读取和写入 json 的一个转换器。Jackson2JsonRedisSerializer 转换器可用于绑定到 Java Bean 或 HashMap 上，可用其替代默认的 JDK 序列化工具。Jackson2JsonRedisSerializer 转换器最大的优点是可以从 Redis 中读取成用户可识别的语言，而不是从 Redis-cli 中读取二进制的乱码。

ObjectMapper 类：可以将 json 格式和 byte 数组来回解析的一个工具，即 json 和 Java Bean 间的映射类。它具有高度的可定制性，可以同时使用不同风格的 json。在 json 对象写出时也可以使用 ObjectWrite 对数据写出。本实例中的 ObjectMapper 相当于给 Jackson2JsonRedisSerializer 转换器设置了一个视图规则。设置自定义的配置 ObjectMapper 是进一步控制 json 序列化的方式，在 ObjectMapper 将 Java Bean 映射成 json 格式后，最终通过 Jackson2JsonRedisSerializer 转换器进行序列化，改变 RedisTemplate 的实现，进而操纵 Java Bean 与 Redis 之间的关系。

StringRedisSerializer 类：将字符串转换为字符节的序列化工具，它使用指定的字符集（默认为 UTF-8）将字符串转换为字符节。通常在与 Redis 的交互中操作 String 时使用。

RedisTemplate.afterPropertiesSet();表示构建了一个新的 RedisTemplate 模型。

9.2.4 编写操作 Redis 的工具类

配置了 RedisTemplate 的相关资源后，可以开始使用 RedisTemplate 带来的函数，其中包括增删改查的一些规则。为了方便使用，可以将 RedisTemplate 整理为 RedisUtils 的一个工具类，会更加方便，工具类整理了判断是否存在该 Key 值、将查询结果放入缓存、删除 Key 值等操作。RedisUtils.java 工具类代码如下。

```java
package com.zfx.util;
import java.util.concurrent.TimeUnit;
import javax.annotation.Resource;
import org.springframework.data.redis.core.RedisTemplate;
import org.springframework.data.redis.core.ValueOperations;
import org.springframework.stereotype.Component;
@Component
public class RedisUtils {

    @Resource(name="redisTemplate")
    private RedisTemplate<String, Object> redisTemplate;

    /**判断是否存在该 Key*/
    public boolean hasKey(String key) {
        return redisTemplate.hasKey(key);
    }
    /**
     * 将查询结果放入缓存
     * 超期时间默认单位为分钟
     * @param key key
     * @param obj value
     * @param time 超期时间
     */
    public void setValue(String key, Object obj, int time) {
        ValueOperations<String, Object> opsForValue = redisTemplate.opsForValue();
        opsForValue.set(key, obj, time, TimeUnit.MINUTES);
    }
    /**
     * 将查询结果放入缓存
     * 超期时间默认 1 小时（60 分钟）
     * @param key key
     * @param obj value
     */
    public void setValue(String key, Object obj) {
        setValue(key, obj, 60);
    }
    /**
     * 删除 key
     * @param key key
     */
    public void deleteKey(String key) {
        redisTemplate.delete(key);
    }
}
```

RedisTemplate 在使用的过程中都需要先从 RedisTemplate 模板获得 ValueOperations

<K,V>对象，其目的是通过 ValueOperations 对象对 Redis 进行操作，其中包含的函数有一系列 set、get、delete 等。

上述代码中对 ValueOperations 接口实现的类（模型对象）还有 DefaultBoundListOperations <K, V>、DefaultSetOperations<K, V>、DefaultBoundZSetOperations <K, V>和 DefaultZSet-Operations<K, V>。Operations 才相当于真正操作了 Redis 缓存。对于使用者来说，只需要通过 RedisTemplate 模板对象获得相应的 Operations 操作模型对象，即可对 Redis 进行操作。

9.2.5 编写实体类及接口调用

编写上述操作 Redis 的 Util 代码后，可直接编写 Java Bean 类及接口，对 RedisUtil 工具类进行测试，创建 Java Bean 实体类 User.java 代码如下。

```java
import java.io.Serializable;

public class User implements Serializable{
    private static final long serialVersionUID = 1L;
    private String username;
    private Integer age;
    //在此省略 setter、getter 函数
}
```

注意，在 Java Bean 中一定要实现 Serializable 接口，因为 Java 应用程序与 Redis 之间是靠 IO 流进行操作的，所以如果没有实现 Serializable 接口，会报相应的错误，其错误明细参见 9.2.8 节。

为方便进行测试，本实例增加了 IndexController.java 类作为 Controller 接口，IndexController.java 代码如下，Controller 接口实现中也只调用了其中一个 set 函数，如果有其他需要，可自行测试函数。

```java
import org.springframework.beans.factory.annotation.Autowired;
import org.springframework.web.bind.annotation.PathVariable;
import org.springframework.web.bind.annotation.RequestMapping;
import org.springframework.web.bind.annotation.RestController;
import com.zfx.entity.User;
import com.zfx.util.RedisUtils;

@RestController
public class IndexController {

    @Autowired
    private RedisUtils redisUtils;

    @RequestMapping("/redisSave/{key}/{value}")
    public String save(@PathVariable String key,@PathVariable String value){
        try {
            redisUtils.setValue(key, new User(value, new Random().nextInt(30)));
            return "success";
```

```
            } catch (Exception e) {
                e.printStackTrace();
                return "error";
            }
        }
    }
```

9.2.6 当前项目结构

本实例通过 RedisTemplate 对象对 Redis 进行了初步的增删改查,并未涉及太多逻辑和函数,当前项目结构如图 9-1 所示。

图 9-1

9.2.7 运行结果

在调用 redisSave 接口对 Redis 进行测试后,可以使用 RedisManager 可视化工具查看刚存入的值,如图 9-2 所示。

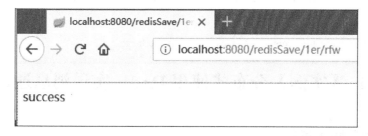

图 9-2

在图 9-2 的接口中任意输入参数,用 RedisManager 读取,如图 9-3 所示。

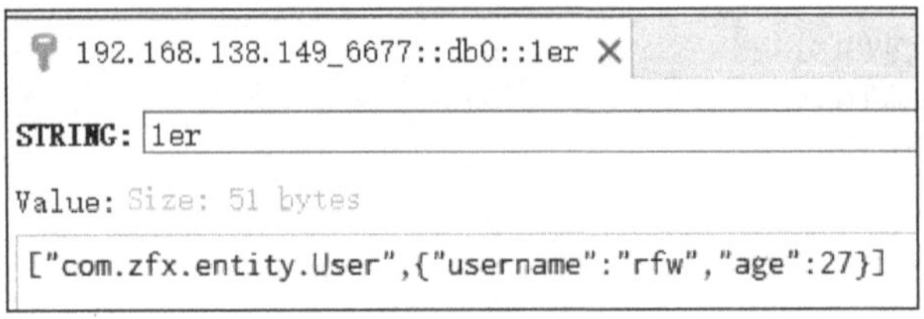

```
┌─────────────────────────────────────────┐
│ 🔑 192.168.138.149_6677::db0::1er  ✕   │
│ STRING: 1er                             │
│ Value: Size: 51 bytes                   │
│ ["com.zfx.entity.User",{"username":"rfw","age":27}] │
└─────────────────────────────────────────┘
```

图 9-3

因为将 RedisTemplate 模板的存入方式重写为以 Jackson2JsonRedisSerializer 类序列化的 json 型格式，而非系统默认的序列化格式，所以在 RedisManager 可视化工具中可以清晰看到存入的内容。下面将使用默认的操作格式操作 Redis，并介绍两者的区别。

用 Spring Boot 微服务通过 Spring Data Redis 的方式操作 Redis 数据库，代码简单，通过实例看到 Spring Data Redis 是如何实现 set 的，以及模板、模型对象之间的流程。在底层中，使用 RedisTemplate 模板操纵 DefaultValueOperations<K, V>模型对象，通过注入 RedisConnectionFactory 连接工厂，使 DefaultValueOperations<K, V>模型对象获得 RedisConnection 连接，以 IO 流的方式操作 Redis。

重写 RedisTemplate 模板在序列化对象时，由 Java 默认的序列化方式改为用 Jackson2JsonRedisSerializer 的序列化方式。通过 ObjectMapper 与 Java Bean 间的映射，由 Java Bean 映射成 json，再由 Jackson2JsonRedisSerializer 进行对 json 的序列化。

9.2.8 实例易错点

1．Java Bean 未增加序列化

Java Bean 未序列化会出现以下异常，此时只要给 Java Bean 增加相应序列化即可。

> No serializer found for class com.zfx.entity.User and no properties discovered to create BeanSerializer (to avoid exception, disable SerializationFeature.FAIL_ON_EMPTY_BEANS)

2．Redis 连接超时

Redis 连接超时会出现以下异常，此时只要检查 Redis 路径端口等相关内容即可。

> Caused by: java.net.SocketTimeoutException: connect timed out

3．传输数据量过大

如果通过 HTTP 向 Java 传输数据，传输的数据过大会出现以下异常，此时只要减少相关存储内容即可。

> java.lang.IllegalArgumentException: Request header is too large

9.3 【实例】分布式使用 Redis 实现消息通信

9.3.1 消息通信应用场景

消息通信常用于异步处理、减少 HTTP 请求压力等相关应用场景，如下所述。

（1）异步处理。例如，注册后通常需给用户发送邮件，在第 6 章中虽然使用了 @Async 注解进行异步处理消息，但是处理消息依旧基于该微服务本身的服务器，消耗其内存等服务器性能，然而通过消息通信的方式可以将异步信息传输给其他服务器，减少服务器压力。

（2）减少 HTTP 请求压力。例如，有业务主动推送信息给用户，类似于给用户推送消息、代金券、广告、站内信等，当一个用户登录后，如果需要给用户实时发送站内信，不可使用前台 Ajax 轮询，因为会造成大量无效的轮询，降低了应用程序的有效使用率。通常会将 Redis 实现发布和订阅的编程内容和 WebSocket 协议联合在一起使用。Redis 消息发布/订阅的方式让信息得以传输到其他相同订阅服务器的频道上，而 WebSocket 负责将收集好的数据推送给前台页面。简而言之，Redis 用于系统层面的实时推送，WebSocket 协议用于系统与页面之间的实时推送。

9.3.2 Redis 与 MQ 一系列消息队列的区别

MQ 一系列消息队列与 Redis 作为消息队列的区别很小，Redis 实时性更高一些。MQ 通常都以 1 秒、1 分、1 小时等作为计时单位进行消息传输，而 Redis 作为高效的缓存服务器，所有数据都在服务器中，它可以实时推送消息给其他机器，不需要设置间隔，只要有数据就会发送。但是 MQ 一系列消息队列会包含完整的消息订阅/发布、Web UI 管理等内容。

Redis 含有持久化功能，Redis 的持久化通常是指对整个 Redis 数据库进行持久化，而 MQ 一系列消息队列的持久化可以选择队列、消息等粒度更小的范围进行持久化。通常 MQ 一系列中间件都有自身的监控 Web 端页面，可以直接看到 MQ 目前的消息数量、被消费的数量、队列的详细情况、CPU 占比、内存的数量等，以方便观察消息队列目前的详细情况。Redis 虽然有 RedisManager 作为可视化管理工具，但是 RedisManager 所提供的能力不够完善，监控和管理能力比一般 MQ 弱得多。

9.3.3 实例背景

本实例创建两个微服务：boot_07_2_send 与 boot_07_2_listener，将通过两个 boot 项目编程。其中，boot_07_2_send 工程负责连接 Redis 后将信息推送出去，boot_07_2_listener 工程在不需要任何额外操作的情况下监听并接收相关信息。

为了完成本实例，首先准备两个空的 Maven 项目，配置 pom.xml 文件代码如下。

```xml
<modelVersion>4.0.0</modelVersion>
<groupId>org.zfx.boot</groupId>
<artifactId>boot_07_2_send</artifactId>
<version>0.0.1-SNAPSHOT</version>

    <parent>
        <groupId>org.springframework.boot</groupId>
        <artifactId>spring-boot-starter-parent</artifactId>
        <version>2.0.4.RELEASE</version>
    </parent>
```

```xml
<dependencies>
    <dependency>
        <groupId>org.springframework.boot</groupId>
        <artifactId>spring-boot-starter-web</artifactId>
    </dependency>
    <dependency>
        <groupId>org.springframework.boot</groupId>
        <artifactId>spring-boot-starter-aop</artifactId>
    </dependency>
    <dependency>
        <groupId>org.springframework.boot</groupId>
        <artifactId>spring-boot-starter-data-redis</artifactId>
    </dependency>
</dependencies>
</project>
```

9.3.4 在 send 微服务中配置模板

在 send 微服务中创建 RedisConfig.java 配置类管理 Redis 模板代码如下。RedisConfig 在 Spring 容器中注册 Redis 的 StringRedisTemplate 模板，用于使 Java 代码读取/操作 Redis 的相关信息。StringRedisTemplate 模板在使用时需要注入 RedisConnectionFactory 连接代码如下。

```java
package com.zfx.config;
import org.springframework.context.annotation.Bean;
import org.springframework.context.annotation.Configuration;
import org.springframework.data.redis.connection.RedisConnectionFactory;
import org.springframework.data.redis.core.StringRedisTemplate;
@Configuration
public class RedisConfig {
    @Bean
    StringRedisTemplate template(RedisConnectionFactory connectionFactory) {
        return new StringRedisTemplate(connectionFactory);
    }
}
```

StringRedisTemplate 类：操作字符串模板，由于对 Redis 队列操作大多基于操作字符串的形式产生，所以 Spring Data Redis 配置了一个专门的类 StringRedisTemplate。也可以使用 RedisTemplate<String,String> 进行操作，因为 StringRedisTemplate 继承了 RedisTemplate<String,String> 类。

9.3.5 在 send 微服务中定时向队列发布数据

在 send 微服务中，通过 Redis 的 Template 模板编写推送消息类 MessageSender.java 代码如下。MessageSender.java 用了任务调度，每 2 秒向 myChannel 渠道中发送一个时间字符串以方便测试。Channel 渠道可以自定义编写，即可以向任意渠道中发送信息，只要订阅者订阅了相同的渠道，就可以实时接收信息。

Redis 的 Channel 渠道可以理解成收音机的频道，有人向该频道发送消息，有人从该频道接收消息。Spring Data Redis 封装了一切复杂的代码，使用简捷，只需要两个类即可完成消息发布的编程内容。

```java
package com.zfx.task;
import org.springframework.beans.factory.annotation.Autowired;
import org.springframework.data.redis.core.StringRedisTemplate;
import org.springframework.scheduling.annotation.EnableScheduling;
import org.springframework.scheduling.annotation.Scheduled;
import org.springframework.stereotype.Component;
@EnableScheduling
@Component
public class MessageSender {

    @Autowired
    private StringRedisTemplate stringRedisTemplate;

    @Scheduled(fixedRate = 2000)
    public void sendMessage(){
        stringRedisTemplate.convertAndSend("myChannel",new SimpleDateFormat("yyyy-MM-dd hh:mm:ss").format(new Date()));
    }
}
```

9.3.6　在 listener 微服务中编写订阅渠道的配置信息

在订阅 Channel 的同时，一旦有消息，会有监听类自动处理，不需要人为干预或轮询等相关操作。

在 listener 微服务中编写 RedisConfig.java 配置类，用于监听 Redis 渠道 Channel 的配置相关代码如下。

```java
package com.zfx.config;
import org.springframework.context.annotation.Bean;
import org.springframework.context.annotation.Configuration;
import org.springframework.data.redis.connection.RedisConnectionFactory;
import org.springframework.data.redis.listener.PatternTopic;
import org.springframework.data.redis.listener.RedisMessageListenerContainer;
import org.springframework.data.redis.listener.adapter.MessageListenerAdapter;
import com.zfx.bean.MessageListener;

@Configuration
public class RedisConfig {

    /**
     * redis 消息监听器容器
```

```java
 * 可以添加多个监听不同话题的 redis 监听器，只需要把消息监听器和相应的消息订阅处理器
绑定，该消息监听器利用反射技术调用消息订阅处理器的相关方法进行业务处理
 */
@Bean
public RedisMessageListenerContainer container(RedisConnectionFactory connectionFactory,
MessageListenerAdapter listenerAdapter) {

    RedisMessageListenerContainer container = new RedisMessageListenerContainer();
    container.setConnectionFactory(connectionFactory);
    //订阅了一个叫 myChannel 的通道
    container.addMessageListener(listenerAdapter, new PatternTopic("myChannel"));
    return container;
}

/**
 * 消息监听器适配器，绑定消息处理器，利用反射技术调用消息处理器的业务方法
 */
@Bean
public MessageListenerAdapter listenerAdapter(MessageListener listener) {
    //给 messageListenerAdapter 传入一个消息接收处理器，利用反射方法调用"receiveMessage"
    return new MessageListenerAdapter(listener, "receiveMessage");
}
}
```

RedisMessageListenerContainer 类：为 Redis 消息监听提供异步行为的容器，处理监听的底层内容。注意，订阅的 Channel 渠道要与发布的 Channel 渠道相同，命名任意。原本只能订阅一个渠道，但是由于 RedisMessageListenerContainer 采用多路复用器的设计模式，所以可以添加多个 MessageListener，即可以重复使用，代码如下。

```
container.addMessageListener(listenerAdapter, new PatternTopic("myChannel"));
```

RedisConnectionFactory 类：目的是获得与 Redis 的 Connection 连接。

MessageListenerAdapter 类：上述代码的核心之一，指消息监听的适配器，可以通过反射的形式将消息托付给 RedisMessageListenerContainer 容器，允许监听的方法以完全异步独立的方式存在。在默认情况下，传入 Redis 消息的内容在传递到目标监听前就被提取，以便让目标方法对消息的内容进行处理。将 MessageListener 注入 MessageListenerAdapter 中后，所有消息都可以从其 receiveMessage 函数中获取。MessageListener 类及其 receiveMessage 函数都由用户自行创建，在 receiveMessage 函数中可读取队列中的相关内容。

PatternTopic 类：相当于渠道的一个载体，渠道名为有参构造。构造 PatternTopic 类后，将 PatternTopic 渠道载体和 MessageListenerAdapter 消息监听适配器，放到 RedisMessage-ListenerContainer 容器中，容器会控制 MessageListenerAdapter 消息监听适配器，将数据放到自定义的 listener 监听实现类中，监听实现类是用户编写的 MessageListener。receiveMessage 是用户监听实现类中的监听函数。

9.3.7 在 listener 微服务中编写监听实现类

因为在监听实现类前依靠 MessageListenerAdapter 构建，直接将实现类放在 MessageListenerAdapter 构造中，所以 MessageListener 监听实现类不需要继承或实现类，只需要编写 MessageListener 监听实现类中的 receiveMessage 函数即可。监听实现类 MessageListener.java 代码如下。

```java
package com.zfx.bean;
import org.springframework.stereotype.Component;
@Component
public class MessageListener {
    /** 接收信息*/
    public void receiveMessage(String message){
        System.out.println("收到一条消息： "+message);
    }
}
```

RedisMessageListenerContainer 容器会自行处理细节，在前面代码中已经配置了渠道名、监听实现类地址、监听函数名等相关内容，Spring Boot 使用 Spring Data Redis 操作 Redis 消息队列（订阅/监听），代码较少。在实际项目中，因业务不同可以在 receiveMessage 函数中对接收的数据进行处理。

9.3.8 当前项目结构

图 9-4 与图 9-5 所示为 Redis 消息发送端与 Redis 消息接收端的项目结构，send 和 listener 的 HTTP 端口不可使用同一个。例如，send 端口为 server.port=8080，listener 端口需改为 server.port=8081，否则报端口冲突错误。

本实例中的 application 资源配置文件与 9.2 节中的 application 资源配置文件相同，在此省略。

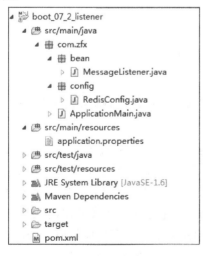

图 9-4   图 9-5

9.3.9　send 微服务与 listener 微服务运行结果

send 微服务与 listener 微服务同时启动后，不需进行任何操作即可从下述日志中观察分布式使用 Redis 实现消息通信的效果。

send：

> 2019-03-05 13:38:25.337　INFO 7380 --- [main] s.b.c.e.t.TomcatEmbeddedServletContainer : Tomcat started on port(s): 8080 (http)
> 2019-03-05 13:38:25.352 INFO 7380 --- [main] com.zfx.ApplicationMain: Started ApplicationMain in 6.89 seconds (JVM running for 7.343)

listener：

> 收到一条消息：2019-03-05 02:41:58
> 收到一条消息：2019-03-05 02:42:00
> 收到一条消息：2019-03-05 02:42:02
> 收到一条消息：2019-03-05 02:42:04

listener 微服务在没有进行任何调用时，每 2 秒会收到一条新的消息，若此时 listener 可以根据收到的相关信息进行判断，有业务则根据消息对业务进行处理。

若想要推送给前台，可以通过 WebSocket 协议向前台进行推送。若想处理相关业务，可以用 json/xml 格式的 String 类型数据进行相互通信。

9.3.10　实例易错点

1. Redis 连接失败

Redis 连接失败通常会报如下异常，需要检查 Redis 地址书写是否有误，或服务器是否开通了相应的端口策略。

> Caused by: io.lettuce.core.RedisConnectionException: Unable to connect to 192.168.138.149:6677

2. Redis 无法接收相关数据

如果无法接收相关数据，需要检查 RedisMessageListenerContainer 配置是否正确，send 和 listener 微服务是否对应同一个 Redis 下的同一个 Channel。

9.4　Spring Cache 与 Spring Data Redis 的区别

Spring 从 3.1 版本开始引入了 Spring Cache 框架。Spring Cache 是基于注解编程的缓存技术，可在自身对 MySQL 进行增删改查的过程中，引入 Spring Cache 注解，自动化处理 MySQL 与 Redis 之间的一致性问题，并且得到相应的返回结果。例如，在 dao 实现类中存储了某个 user，使用 Spring Cache 的相关注解后，dao 实现类不仅存储在 MySQL 中，也会存储并更新到 Redis 上，待下次查询时，优先查询 Redis 中是否有相应数值，若存在则直接返回，若不存在则到 MySQL 中查询，极大化地减小了 MySQL 的并发压力。如果所有缓存都先 if 判断，再编写各种代码，代码的重复率高，而且耦合度也高，不方便维护与修改。如果使用 Spring Cache 注解，就不需要编写 if 等相关代码了。

Spring Data Redis 依靠 RedisTemplate 操作 Redis 数据库，而 Spring Cache 依靠注解操作 Redis 数据库。而 RedisTemplate 实现都是用户手写的，对条件查询可以整理，但如果项目过大，耦合性可能更高，日后维护也会变得更加困难。

Spring Cache 对条件查询和分页查询等批量产生的情况处理起来十分烦琐。例如，要查询 MySQL 第 2 页的 20 条数据，但因为后来修改了 ID=5 的数据，ID=5 不一定在第 2 页，所以可能导致 Redis 和 MySQL 的脏读现象，即数据的一致性遭到了污染。如果查询的某个 list 会经常变更，就不要使用 Reids 进行存储，进而直接从 MySQL 中读取；对于 list 不经常变更的情况，使用 Spring Cache 进行增删改后，删除所有关于 list 的缓存，再进行 Redis 存储，Spring Cache 更适用于变更较少却调用频繁的情况。

9.5 【实例】保持 MySQL 与 Redis 数据一致性

9.5.1 实例背景

本实例创建 boot_07_3 工程，将使用 Spring Boot 项目整合 Spring Data JPA 和 Spring Cache，分别操作 Redis 和 MySQL，但要具有如下功能。

（1）在查询 MySQL 前，优先查询 Redis，若 Redis 中没有，再去查询 MySQL。

（2）在查询 MySQL 前，优先查询 Redis，若 Redis 中有，直接依靠 Redis 返回。

（3）若 Redis 中有某一数值，而对 MySQL 中该数值进行更改时，要保证 Redis 和 MySQL 之间的数据一致性。

先准备一个空的 Maven 项目，并且编辑它的 pom.xml 和 Spring Boot 启动类代码。pom.xml 部分代码如下。

```xml
<parent>
    <groupId>org.springframework.boot</groupId>
    <artifactId>spring-boot-starter-parent</artifactId>
    <version>2.0.4.RELEASE</version>
</parent>

<dependencies>
    <dependency>
        <groupId>org.springframework.boot</groupId>
        <artifactId>spring-boot-starter-data-jpa</artifactId>
    </dependency>
    <!-- 整合 redis -->
    <dependency>
        <groupId>org.springframework.boot</groupId>
        <artifactId>spring-boot-starter-data-redis</artifactId>
    </dependency>
```

```xml
        <dependency>
            <groupId>mysql</groupId>
            <artifactId>mysql-connector-java</artifactId>
        </dependency>
        <dependency>
            <groupId>org.springframework.boot</groupId>
            <artifactId>spring-boot-starter-web</artifactId>
        </dependency>
    </dependencies>
```

9.5.2 编写资源配置文件

将 Redis 数据库参数和 MySQL 数据库参数写到 application.properties 资源配置文件中，代码如下。

```
server.port=8080
spring.jpa.database=mysql
spring.jpa.show-sql=true
spring.jpa.hibernate.ddl-auto=update
spring.jackson.serialization.fail-on-empty-beans=false
spring.datasource.url=jdbc:mysql://192.168.138.149:3306/springboot-redis?useUnicode=true&characterEncoding=utf-8&allowMultiQueries=true&useSSL=false
spring.datasource.username=root
spring.datasource.password=mypassword
spring.datasource.driver-class-name=com.mysql.jdbc.Driver
spring.redis.database=0
spring.redis.host=192.168.138.149
spring.redis.port=6379
spring.redis.password=
spring.redis.jedis.pool.max-active=10
spring.redis.jedis.pool.max-wait=-1
spring.redis.jedis.pool.max-idle=8
spring.redis.jedis.pool.min-idle=1
```

9.5.3 编写实体类 Java Bean

与 7.4 节相同，编写 BookEntity.java 实体类 Java Bean，用于 JPA 映射数据库。BookEntity.java 实体类代码如下。

```java
import javax.persistence.Column;
import javax.persistence.Entity;
import javax.persistence.GeneratedValue;
import javax.persistence.GenerationType;
import javax.persistence.Id;
import javax.persistence.Table;
import org.hibernate.annotations.Proxy;
@Entity
@Table(name="book")
```

```
@Proxy(lazy = false)//注意，此注解一定要加，否则在查询后，再次从缓存中取值会报错
public class BookEntity implements Serializable{

    private static final long serialVersionUID = 1L;
    @Id//主键
    @GeneratedValue(strategy = GenerationType.IDENTITY)//主键由数据库自动生成（主要是自动增长型）
    private Long id;
    @Column(name="book_name")
    private String bookName;
    @Column(name="book_author")
    private String bookAuthor;
    @Column(name="book_count")
    private Integer bookCount;
//省略 setter、getter。也可以用@Data
}
```

9.5.4 编写 JPA 仓库

编写 BookRepository.java 仓库类代码如下。因为本实例只用 JPA 仓库中较简单的函数，所以仓库只继承了 JpaRepository<BookEntity,long>，并没有声明其他抽象函数。

```
import org.springframework.data.jpa.repository.JpaRepository;
import com.zfx.entity.BookEntity;
public interface BookRepository extends JpaRepository<BookEntity, Long>{}
```

9.5.5 编写 Service 接口及实现类

编写 BookService.java 接口类代码如下，自定义部分增删改查操作。

```
public interface BookService {
    //更新或保存
    BookEntity save(BookEntity book);

    BookEntity getBookById(Long id);

    List<BookEntity> selectAll();

    void deleteById(Long id);
}
```

编写 BookServiceImpl.java 实现类代码如下。在调用 JPA 函数进行增删改查的同时，使用 Spring Cache 的注解修饰每个 Service 函数，以达到保持 MySQL 与 Redis 数据一致性的目的。

```
package com.zfx.service.impl;
import java.util.List;
import org.springframework.beans.factory.annotation.Autowired;
import org.springframework.cache.annotation.CacheConfig;
```

```java
import org.springframework.cache.annotation.CacheEvict;
import org.springframework.cache.annotation.CachePut;
import org.springframework.cache.annotation.Cacheable;
import org.springframework.stereotype.Service;
import com.zfx.entity.BookEntity;
import com.zfx.repository.BookRepository;
import com.zfx.service.BookService;

@Service
@CacheConfig(cacheNames = "book")
public class BookServiceImpl implements BookService{
    @Autowired
    private BookRepository bookRepository;

    @CachePut(key = "#p0.id")
    public BookEntity save(BookEntity book) {
        BookEntity bookEntity = bookRepository.save(book);
        return bookEntity;
    }
    @Cacheable(key = "#p0")
    public BookEntity getBookById(Long id) {
        BookEntity bookEntity = bookRepository.getOne(id);
        return bookEntity;
    }
    @Cacheable(key="#root.methodName")
    public List<BookEntity> selectAll() {
        List<BookEntity> findAll = bookRepository.findAll();
        return findAll;
    }
    @CacheEvict(key = "#p0")
    public void deleteById(Long id) {
        BookEntity bookEntity = new BookEntity();
        bookEntity.setId(id);
        bookRepository.delete(bookEntity);
    }
}
```

@Cacheable 注解：@Cacheable 的作用主要是针对方法进行配置，能根据方法的请求参数对其进行缓存。每调一次该方法，都会进行相应的缓存行为。通常在读取 MySQL 时使用@Cacheable 注解。

@Cacheable 注解内部参数 Key：Key 是 K/V@Cacheable 中最重要的一部分，它代表了相应内容存入 Redis 时的 Key 值，以达成 Redis 的 K/V 键值对格式。它可以 String 形式存在，也可以用 SPEL（用于动态计算秘钥的 Spring 表达式语言）进行存储。除 Key 参数外，还有 condition 条件参数、cacheNames 命名空间等参数。

@CachePut 注解：对于使用@Cacheable 标注的方法，Spring 在每次执行前都会检查

Cache 中是否存在相同 Key 的缓存元素，如果存在就不执行该方法，而直接从缓存中获取结果并返回，否则执行该方法并将返回结果存入指定的缓存中。@CachePut 也可以声明一个方法支持缓存功能。与@Cacheable 不同的是，使用@CachePut 标准的方法在执行前，不会强制检查缓存中是否存在之前执行过的结果，而是每次都会执行该方法，并将执行结果以键值对的形式存入指定的缓存中。

@CacheEvict 注解：与@CachePut 刚好相反，@CacheEvict 为删除，一般增删改 MySQL 的语句都可以使用@CacheEvict，在更改 MySQL 后，@CacheEvict 注解也会修改 Redis 内的数值。

@CacheConfig 注解：@CacheConfig 注解是 cacheNames、keyGenerator、cacheManager、cacheResover 的组合注解，提供了缓存相关配置的共享机制。当@CacheConfig 注释出现在一个类上时，它提供了许多定义缓存的默认配置。

@CacheConfig 注解内 cacheNames：为定义的缓存操作考虑的默认缓存名称，相当于一个命名空间，可以修改某命名空间下的配置信息或使用某命名空间下的配置信息。

@CacheConfig 注解内 keyGenerator：缓存密钥生成器，用来在某个函数上创建密钥。keyGenerator 接口下的实现方式有 DefaultKeyGenerator、KeyGeneratorAdpter、SimpleKeyGenerator，不同的实现方式只是对密钥生成策略的不同，用户也可以自行实现 KeyGenerator 接口。keyGenerator 用不同的策略与方式生成 Key 值，不需要考虑 Key 的样式，但是不建议滥用，因为可能出现一些出乎意料的情况。

@CacheConfig 注解内 cacheManager：管理缓存中的数据，相当于一个缓存管理器。可以理解为 Spring Cache 连接 Redis 的 Connection 连接被存储在@CacheConfig 中，还可以存储多个链接或管理命名空间下的 Cache 数据。

@CacheConfig 注解内 cacheResover：相当于 Spring Cache 中的上下文。

#p0 表达式：#p0 指第一个入参值作为 Key，#p0.id 指用第一个入参对象中的 ID 参数作为 Key。由此延伸，#p1、#p1.name 都是可以理解的。也可以直接强写名字如下述代码：

```
@Cacheable(key="T(String).valueOf(#name).concat('-').concat(#password)")
public void method(int name,String password){
    //上述代码拼接了一个 Key 值
}
```

#root.methodName 表达式：SpringEL 可以用 root 对象生成 Key 值，示例如表 9-1 所示。

表 9-1　SpringEL 的 root 对象示例

属性名称	描述	表达式示例
methodName	当前方法名	#root.methodName
method	当前方法	#root.method.name
target	当前被调用的对象	#root.target
targetClass	当前被调用的对象的 class	#root.targetClass
args	当前方法参数组成的数组	#root.args[0]
caches	当前被调用的方法使用的 Cache	#root.caches[0].name

9.5.6 编写 Controller 接口进行测试

为了方便测试，编写 BookController.java 类提供相关接口，其 Controller 只编写了一个测试的代码进行增删改查，代码如下。

```java
import org.springframework.beans.factory.annotation.Autowired;
import org.springframework.web.bind.annotation.RequestBody;
import org.springframework.web.bind.annotation.RequestMapping;
import org.springframework.web.bind.annotation.RestController;
import com.zfx.entity.BookEntity;
import com.zfx.service.BookService;
@RestController
@RequestMapping("book")
public class BookController {

    @Autowired
    private BookService bookService;

    @RequestMapping("/save")
    public BookEntity save(@RequestBody BookEntity book) {
        return bookService.save(book);
    }
    @RequestMapping("/getById")
    public BookEntity getById(Long id) {
        BookEntity book = bookService.getBookById(id);
        return book;
    }
    @RequestMapping("/selectAll")
    public List<BookEntity> selectAll(){
        return bookService.selectAll();
    }
}
```

9.5.7 当前项目结构

如图 9-6 所示，因为 Spring Cache 与代码完美融入，几乎看不到整合 Redis 的类，即直接通过注解嵌入程序中，使项目整体更加简捷。

9.5.8 运行结果

（1）先用 postman 调用 /book/selectAll 接口，如图 9-7 所示，/book/selectAll 接口返回的是空，因为此刻 MySQL 中并没有数据。

图 9-6

图 9-7

因此显示的是正确的，同时在查询所有信息后，如图 9-8 所示，检查 redisManager 管理器可以发现@Cacheable 注解在查询后返回结果为空时，并不会将空的值存入 Redis 中。

（2）如果存储后，调用 selectAll 或 getId 接口，检查 redisManager 管理器可以发现 Reids 数据库已经存储了相应的数值，下次读取时直接从 Redis 缓存中读取相应数值，如图 9-9 所示。

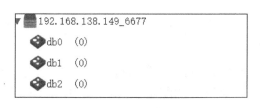

图 9-8

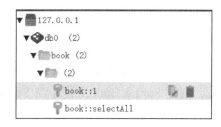

图 9-9

9.5.9 实例易错点

1. MySQL 主键错误相关

由于 ID 字段在数据库表中作为主键，但在 Java 代码插入的过程中，没有给 ID 赋值，且没有让其自增长，所以会报出如下异常。

java.sql.SQLException: Field 'id' doesn't have a default value

解决方法是修改数据库表中的 ID 属性，让其自增，并且在 Java 代码插入的过程中，不插入 ID 数据，即使数据库的 ID 自增长也不会报错。

2. 微服务实体类映射错误

Java Bean 与 MySQL 互相映射期间若无法相互对应，则会报如下异常。

java.lang.IllegalArgumentException: Null key returned for cache operation (maybe you are using named params on classes without debug info?) Builder[public com.zfx.entity.BookEntity com.zfx.service.impl.

```
BookServiceImpl.getBookById(java.lang.Long)] caches=[book] | key='#p0' | keyGenerator='' | cacheManager=''
| cacheResolver='' | condition='' | unless='' | sync='false'
```

此时应检查 MySQL 与 Java Bean 互相映射的内容是否有误。

3. Cache 的 SpringEL 表达式编写错误

若 ServiceImpl 中有关于 Cache 注解内部的 SpringEL 表达式编写错误，则会报以下异常。

```
org.springframework.expression.spel.SpelParseException: Expression [#123123p0] @1: EL1043E:
Unexpected token. Expected 'identifier' but was 'literal_int'
```

此时应检查相关 SpringEL 表达式是否编写错误。

9.6 本章小结

  Redis 是最简洁的非关系型缓存数据库，在实际工作中使用较多。9.2 节与 9.6 节分别使用 Spring Data Redis 与 Spring Cache 整合 Redis 以实现缓存的增删改查操作。
  9.5 节使用 Redis 作为消息通信中间件进行多个微服务之间的通信。

9.7 习　　题

  1. 创建 Spring Boot 工程，使用 Spring Data Redis 整合 Redis 非关系型数据库，并进行增删改查操作，记录 Redis 可视化软件中的相关结果。

  2. 创建 Spring Boot 工程 1 与 Spring Boot 工程 2，两个工程通过 Redis 达到消息通信的目的，即在 Spring Boot 工程 1 并不执行任何命令和函数的同时，Spring Boot 工程 2 向 Spring Boot 工程 1 中传输数据，Spring Boot 工程 1 能正确收到。

  3. 创建 Spring Boot 工程，整合 Spring Data JPA 和 Spring Cache 分别用来操作 Redis 与 MySQL，达到以下目的。

  （1）在查询 MySQL 前，优先查询 Redis，若 Redis 中没有，再查询 MySQL。

  （2）在查询 MySQL 前，优先查询 Redis，若 Redis 中有，直接使用 Redis 进行返回。

  （3）若 Redis 中有某一数值，在对 MySQL 中该数值进行更改时，要保证 Redis 和 MySQL 之间的数据一致性。

第 10 章 微服务的任务调度与分布式的任务调度

任务调度是指使程序在某个时间点做某件事,或者每隔某个时间点做某件事。例如,银行工作中有一个"跑批任务",银行核心系统每天六点会与金融理财系统进行对账,双方各提交一份 txt 文件并查询两份文件是否有不一致之处,该业务要求的时间点为每天六点,即是靠任务调度来实现的。

10.1 【实例】微服务整合任务调度

10.1.1 实例背景

本实例将创建 boot_04_1 工程,编写每隔几秒可以自动执行一个函数的程序。通过应用程序介绍@EnableScheduling 注解和@Scheduled 注解,初步认识任务调度及其使用方法。

10.1.2 编写任务调度实现类

编写任务调度的实现类 TaskScheduled.java 代码如下。此处要注意的是 @EnableScheduling 注解用于任务调度的启动,不可遗漏,否则任务调度无法正常启动。

@Scheduled()注解内部使用的任务调度是 cron 表达式,cron 表达式的 Value 值可以直接使用【$】符号在 application.properties 等文件中获取。

```
package com.zfx.task;
import java.text.SimpleDateFormat;
import java.util.Date;
import org.springframework.scheduling.annotation.EnableScheduling;
import org.springframework.scheduling.annotation.Scheduled;
import org.springframework.stereotype.Component;
@Component
@EnableScheduling
public class TaskScheduled {

    private static final SimpleDateFormat dateFormat = new SimpleDateFormat("HH:mm:ss");

    @Scheduled(fixedDelayString = "${jobs.fixedDelay}")
    public void getTaskA() {
    System.out.println("任务 A,当前时间: " + dateFormat.format(new Date()));
```

```
    }

    @Scheduled(cron = "${jobs.cron}")
    public void getTaskB() {
        System.out.println("任务 B，当前时间： " + dateFormat.format(new Date()));
    }
}
```

10.1.3 编写资源配置文件

编写 application.properties 资源配置文件中的相关参数如下。其中关于 Job 的 Key 值都是用户自行定义的，由 SpringEL 表达式${jobs.cron}与${jobs.fixedDelay}取出即可。

```
jobs.fixedDelay=5000
jobs.cron=0/5 * * * * ?
```

10.1.4 当前项目结构

当前任务调度的应用程序项目结构如图 10-1 所示，可达到定时执行任务的效果。

10.1.5 运行效果

执行程序后不需要任何操作，@Scheduled 注解修饰的函数就会定时执行，如图 10-2 所示。

图 10-1　　　　　　　　　　图 10-2

10.1.6 实例易错点

（1）@EnableScheduling 注解：用于任务调度的启动。若整个 Spring Boot 应用程序中都不含有@EnableScheduling 注解，在 Spring Boot 应用程序执行 ApplicationMain 方法时不会报错，但是不会执行该任务调度。

（2）@EnableScheduling 注解中虽然引入了 SchedulingConfiguration.class 类，并且 SchedulingConfiguration 类中涵盖@Configuration 函数，但是由于@EnableScheduling 注解没有直接被@Configuration 注解所修饰，所以@EnableScheduling 注解并不含有相关 Spring 注册的注解，如 @Service、@Controller、@Configuration、@Component 等，即 @EnableScheduling 注解不会直接在 Spring 容器中注册 Bean，需要在类名前增加相应的注册注解。

（3）@EnableScheduling 可以写在 ApplicationMain.class 类头上。

10.2 @Scheduled 注解详解

@Scheduled 注解的参数及释义如表 10-1 所示。

表 10-1 @Scheduled 注解的参数及释义

参 数	释 义	示 例
cron	需要输入一个 cron 表达式，根据表达式对任务调度的时间进行控制，返回一个可以解析为 CRON 计划的表达式	@Scheduled(cron = "0/5 * * * * ?")
fixedDelay	输入一个 long 值，从上一轮结束到下一轮开始之间相隔 long 毫秒，如上一轮结束到下一轮开始相隔 10 000 毫秒	@Scheduled(fixedDelay = 10000)
fixedDelayString	与 fixedDelay 意义相同，但是输入参数改成 String	@Scheduled(fixedDelayString = "10000")
fixedRate	输入一个 long 值，从上一轮开始到下一轮开始相隔 long 毫秒，如上一轮开始到下一轮开始相隔 5000 毫秒	@Scheduled(fixedRate = 5000)
fixedRateString	与 fixedRate 意义相同，但是输入参数改成 String	@Scheduled(fixedRate = "5000")
initialDelay	输入一个 long 值，初始化 long 毫秒后才进行任务调度，如 Spring 容器在初始化 1000 毫秒后进行第一次任务调度，第一次任务调度开始时间隔 5000 毫秒进行第二次任务调度	@Scheduled(initialDelay = 1000，fixedRate = 5000)
initialDelayString	与 initialDelay 意义相同，但是输入参数改成 String	@Scheduled(initialDelayString = "1000"，fixedRate = 5000)

10.2.1 cron 表达式

cron 完整表达式（0/5 * * * * ?）如下。

{秒} {分} {小时} {日期} {月份} {星期} {年份(可为空)}

cron 表达式示例如下。

"0 0 15 ? * WED"

上式表示在每星期三 15:00 执行（年份通常省略）。

10.2.2 每个字段允许值

允许值表示允许输入的值范围，cron 表达式除正常数值型的字符外，还允许输入 cron 表达式允许的特殊字符，如表 10-2 所示。

表 10-2 cron 表达式允许的特殊字符

字　　段	允　许　值	允许的特殊字符
秒（Second）	0~59 的整数	, - * /
分（Minute）	0~59 的整数	, - * /
小时（Hour）	0~23 的整数	, - * /
日期（Day of Month）	1~31 的整数（需要考虑每月的天数）	, - * ? / L W C
月份（Month）	1~12 的整数或 JAN, FEB, MAR, APR, MAY, JUN, JUL, AUG, SEP, OCT, NOV, DEC	, - * /
星期（Day of Week）	1~7 的整数或 SUN, MON, TUE, WED, THU, FRI, SAT（1=SUN）	, - * ? / L C #
年份(可选，留空)（Year）	1970~2099	, - * /

10.2.3 cron 特殊字符意义

表 10-3 所示为 cron 表达式允许的特殊字符及意义。

表 10-3 cron 表达式允许的特殊字符及意义

特殊字符	意　　义
,	用于将多种情况间隔开（如果写成 5,20,35 表示该情况下才执行，如 5,20,35 * * * * ? 表示每分钟的第 5, 20, 35 秒时才会执行）
-	表示范围，a-b 表示从 a 到 b 的范围（例如，5-20 * * * * ? 表示每分钟的 5~20 秒，每秒执行一次；5-20/5 * * * * ? 表示每分钟的 5~20 秒内，每 5 分钟执行一次）
*	表示所有值中的每个
/	分割符，分为前后两个值，a/b 表示从 a 开始，之后每隔 b 执行一次
?	只能用于日期和星期，不管几日或星期几，表示所有值中的某一个值。注意，日期和星期不能同时使用?，日期和星期两个值必须有一个是指定的
L	只能用于日期和星期，注意日期与星期不能同时使用；在日期上使用表示每月的最后一天（如 0 0 15 L * ? 表示每个月最后一天的下午 3 点执行），在星期上使用表示最后一天即 7 或 SAT（周六）；如果与数字一起使用，表示每月最后一个星期几（如 0 0 15 ? * 6L 表示每个月最后一个星期五的下午 3 点执行）
W	只能用于日期，表示工作日，即周一至周五，需要与数字一起使用，该值存在就近原则（例如，0 0 15 16W ? * 表示每个月 16 日工作日的下午 3 点执行，如果 16 日是周六，则 15 日周五执行；如果 16 日为周日，则 17 日周一执行）。注意，该用法只会在当前月计算值，不会越过当前月，LW 表示最后一个工作日
#	只能用于星期，a#b 表示该月第 b 周的星期 a，如果指定的日期不存在，就不会执行

10.2.4 常用 cron 表达式

由于 cron 表达式编写稍有复杂之处，所以表 10-4 中整理了常用的 cron 表达式，cron 表达式与正则表达式有些类似，用相应规定好的数字及符号表达 cron 表达式的意义，即以什么频率和条件执行当前任务。

表 10-4　常用的 cron 表达式

cron 表达式	释　　义
0 15 10 ? * *	每天上午 10:15 触发
0 15 10 * * ?	每天上午 10:15 触发
0 15 10 * * ? *	每天上午 10:15 触发
0 15 10 * * ? 2005	2005 年的每天上午 10:15 触发
0 * 14 * * ?	在每天下午 2 点到下午 2:59 期间每 1 分钟触发
0 0/5 14 * * ?	在每天下午 2 点到下午 2:55 期间每 5 分钟触发
0 0/5 14,18 * * ?	在每天下午 2 点到 2:55 期间和下午 6 点到 6:55 期间每 5 分钟触发
0 0-5 14 * * ?	在每天下午 2 点到下午 2:05 期间每 1 分钟触发
0 10,44 14 ? 3 WED	每年三月的星期三下午 2:10 和 2:44 触发
0 15 10 ? * MON-FRI	周一至周五的上午 10:15 触发
0 15 10 15 * ?	每月 15 日上午 10:15 触发
0 15 10 L * ?	每月最后一日的上午 10:15 触发
0 15 10 ? * 6L	每月的最后一个星期五上午 10:15 触发
0 15 10 ? * 6L 2002-2005	2002 年至 2005 年每月的最后一个星期五上午 10:15 触发
0 15 10 ? * 6#3	每月的第三个星期五上午 10:15 触发
0 15 10 L * ? 2002-2005	2002 年至 2005 年的每月最后一天上午 10:15 触发
0 15 10 LW * ? 2002-2005	2002 年至 2005 年的每月最后一个工作日上午 10:15 触发
0 15 10 5W * ? 2002-2005	2002 年至 2005 年的每月离 5 号最近的工作日上午 10:15 触发，若 5 号是工作日，则在 5 号触发；若 5 号是周六，则在 4 号触发；若 5 号是周日，则在 6 号触发

10.3　任务调度的分布式

单点情况下的任务调度一旦出现服务器异常，就无法正常执行已规定的任务，此时需要依靠任务调度分布式的方案解决，但是任务调度无法使用 Ribbon 或 Nginx 进行客户端或服务器的高可用和负载均衡，因为如果用了以上两种方式进行任务调度，同时启动多台服务器将在规定的时间内执行相同的任务，造成任务的重复执行、数据库脏读等相关问题。

10.3.1　任务调度的分布式解决方案

业内任务调度的分布式解决方案有很多种，如 Opencron、LTS、XXL-Job、Elastic-Job、Uncode-Schedule、Antares、SkySchedule、timerTask 和 Quartz。因为任务调度的分布式原理较为简单，所以出现了各式各样的任务调度分布式框架，本书推荐 Opencron、LTS、XXL-Job、Elastic-Job 等拥有自身 UI 的任务调度分布式框架或 Quartz 传统高效的任务调度分布式框架。

10.3.2 任务调度的分布式实现原理

任务调度的分布式实现原理类似于事务的分布式实现原理，即将任务调度的详细内容存储在某一介质（如 MySQL）中，多台服务器同时连接 MySQL 数据库，得知当前任务是否有节点正在执行，若就近有节点正在执行，其他服务器不会重复执行。

此处以 Quartz 举例。集群仅适用 Quartz 框架内的 JDBC-Jobstore 调度、JobStoreTX 调度、JobStoreCMT 调度，并且基本上通过使集群的每个节点共享相同的数据库来工作。集群的每个节点都尽可能快地触发作业。当 Triggers 的触发时间出现时，获取它的第一个节点（伪随机获取）是将触发它的节点（通过在其上放置一个锁定状态）。负载平衡机制对于繁忙的调度器（如大量的 Triggers）是近似随机的，但是对于非忙（如很少的 Triggers）调度器而言，有利于同一个节点。当其中一个节点在执行一个或多个作业期间失败时，发生故障切换。当节点出现故障时，其他节点会检测到该状况并识别数据库中在故障节点内正在进行的作业。任何标记为恢复的作业（在 JobDetail 上都具有"请求恢复"属性）将被剩余的节点重新执行。没有标记为恢复的作业将在下一次相关的 Triggers 触发时简单地被释放得以执行。

10.4 【实例】微服务整合任务调度分布式

10.4.1 实例背景

本实例创建 boot_async_00 工程与 boot_async_00-2 工程。boot_async_00 工程与 boot_async_00-2 工程除端口号、项目名称不同外，核心代码基本一致，都将通过微服务整合 Quartz 任务调度分布式框架，展示在分布式情况下如何进行任务调度管理，测试时将同时运行 boot_async_00 工程与 boot_async_00-2 工程，并保证其中只有一个任务调度处在正常执行状态，而另一个工程不会执行任务调度的任何操作。

10.4.2 增加 Quartz 依赖

在 pom.xml 文件中只需要增加如下代码即可集成 Quartz。

```
<dependency>
    <groupId>org.springframework.boot</groupId>
    <artifactId>spring-boot-starter-quartz</artifactId>
</dependency>
```

10.4.3 在数据库中增加 Quartz 分布式的管理表

大部分任务调度框架都依靠相关介质的存储，使不同的服务器访问相同介质，从该介质中得到相关的调度信息，根据调度信息执行相关的调度任务，以此来达到分布式的效果。

Quartz 依靠 MySQL 作为任务调度管理分布式数据源的介质，需要下载相关的管理表，将相关的管理表一键导入 MySQL，效果如图 10-3 所示。

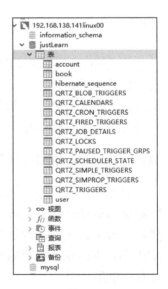

图 10-3

Quartz 的 SQL 文件目录如图 10-4 所示。

quartz / quartz-core / src / main / resources / org / quartz / impl / **jdbcjobstore** /

zemian improv: Use username quartz as init liquibase		Latest commit 0f82063 on 1 Mar
..		
liquibase.quartz.init.xml	improv: Use username quartz as init liquibase	6 months ago
tables_cloudscape.sql	Move database schema scripts to quartz-core	3 years ago
tables_cubrid.sql	Move database schema scripts to quartz-core	3 years ago
tables_db2.sql	Move database schema scripts to quartz-core	3 years ago
tables_db2_v72.sql	Move database schema scripts to quartz-core	3 years ago
tables_db2_v8.sql	Move database schema scripts to quartz-core	3 years ago
tables_db2_v95.sql	Move database schema scripts to quartz-core	3 years ago
tables_derby.sql	Move database schema scripts to quartz-core	3 years ago
tables_derby_previous.sql	Move database schema scripts to quartz-core	3 years ago
tables_firebird.sql	Move database schema scripts to quartz-core	3 years ago
tables_h2.sql	Move database schema scripts to quartz-core	3 years ago
tables_hsqldb.sql	Merge remote-tracking branch 'origin/quartz-2.2.x'	2 years ago
tables_hsqldb_old.sql	Move database schema scripts to quartz-core	3 years ago
tables_informix.sql	Move database schema scripts to quartz-core	3 years ago
tables_mysql.sql	Move database schema scripts to quartz-core	3 years ago
tables_mysql_innodb.sql	Move database schema scripts to quartz-core	3 years ago
tables_oracle.sql	Move database schema scripts to quartz-core	3 years ago
tables_pointbase.sql	Move database schema scripts to quartz-core	3 years ago
tables_postgres.sql	Merge remote-tracking branch 'origin/quartz-2.2.x' into zemian_160-po...	2 years ago
tables_sapdb.sql	Move database schema scripts to quartz-core	3 years ago
tables_solid.sql	Move database schema scripts to quartz-core	3 years ago
tables_sqlServer.sql	Add missing foreign key for BLOB triggers table for ms sql server #76	2 years ago
tables_sybase.sql	Based on #245 fix, we improve related varchar length	6 months ago

图 10-4

Quartz 对接 MySQL 通常使用 tables_mysql_innodb.sql 文件。

10.4.4　编写资源配置文件

Quartz 在 application.yml 中增加的资源配置文件代码如下。其中，spring.quartz.job-store-type 参数为 Quartz 访问数据库的交互方式，与当前程序中 spring-data-jpa 的执行方式无任何关系。因此，可以让 Quartz 使用 JDBC 的方式访问 MySQL 数据源，获得分布式的相关配置表。

```
server:
  port: 8081
spring:
  datasource: # 数据库配置
    driver-class-name: com.mysql.jdbc.Driver
    url: jdbc:mysql://192.168.138.141:3306/justLearn
    username: zhang
    password: zhang55614233
  jpa: # JPA 配置
    database: mysql
    show-sql: true
  quartz:
```

```yaml
# quartz 相关属性配置
  properties:
    org:
      quartz:
        scheduler:
          instanceName: clusteredScheduler
          instanceId: AUTO
        jobStore:
          class: org.quartz.impl.jdbcjobstore.JobStoreTX
          driverDelegateClass: org.quartz.impl.jdbcjobstore.StdJDBCDelegate
          tablePrefix: QRTZ_
          isClustered: true
          clusterCheckinInterval: 10000
          useProperties: false
        threadPool:
          class: org.quartz.simpl.SimpleThreadPool
          threadCount: 10 # 开启 10 个线程
          threadPriority: 5
          threadsInheritContextClassLoaderOfInitializingThread: true
  job-store-type: jdbc
```

10.4.5 创建任务调度管理 Java Bean

创建任务调度配置的 Java Bean 代码如下。该 Java Bean 不需要对应数据库中存储的 Quartz 管理分布式的 MySQL 表，但会管理关于任务的执行与生成。MySQL 中存储有关 Quartz 的表都将由 Quartz 自行管理，通常情况下不需要用户对其进行增删改查操作。

```java
package com.zfx.task.bean;
import java.util.Map;
import org.quartz.Job;
import org.quartz.JobKey;

//@Entity
public class TaskBean {

    /**定时任务的名字和分组名  JobKey*/
    private JobKey jobKey;
    /**定时任务的描述(可为定时任务本身的描述,也可以是触发器的)*/
    private String description;
    /**定时任务的执行 cron*/
    private String cronExpression;
    /**定时任务的元数据*/
    private Map<?, ?> jobDataMap;
    /**定时任务的具体执行逻辑类*/
    private Class<? extends Job> jobClass;
    //以下省略 setter、getter 方法

}
```

10.4.6 创建所需执行的任务

创建 TaskJob.java 类继承 Quartz 的 Job 类并重写 execute 函数，以用来创建所需执行的任务，该类为实际执行任务的类，内部编写所需执行的任务逻辑即可，Quartz 会根据相关配置执行该 Job，其代码如下。

```
package com.zfx.task;
import java.text.SimpleDateFormat;
import java.util.Date;
import org.quartz.DisallowConcurrentExecution;
import org.quartz.Job;
import org.quartz.JobExecutionContext;
import org.quartz.JobExecutionException;

@DisallowConcurrentExecution
public class TaskJob implements Job{

    @Override
    public void execute(JobExecutionContext arg0) throws JobExecutionException {
        System.out.println("TaskService,执行: "+new SimpleDateFormat().format(new Date()));
    }
}
```

@DisallowConcurrentExecution 注解：将修饰的类标记为含有独一无二 JobKey 和 JobDetail 的类，该类不能被多个类并发执行。

Job：Job 易实现，在接口中只有一个 execute 方法，实现该方法后可以指定任务调度执行的任务 Job。

Job Key：键包括名称和组，名称在 Job Group 组内必须是唯一的。若只指定一个名称，则默认组将使用 name。Job Key 键与 Job Group 指定了该 Job 的唯一性。

10.4.7 创建执行任务的操作类

创建 TaskService.java 类，通过启动、暂停、恢复、获取 Job 的相关信息，修改 cron 表达式等方式操作 Quartz 相关的 Job，代码如下。

```
package com.zfx.task.service;
import java.util.Map;
import org.quartz.CronExpression;
import org.quartz.CronScheduleBuilder;
import org.quartz.CronTrigger;
import org.quartz.Job;
import org.quartz.JobBuilder;
import org.quartz.JobDataMap;
import org.quartz.JobDetail;
import org.quartz.JobKey;
import org.quartz.Scheduler;
```

```java
import org.quartz.SchedulerException;
import org.quartz.Trigger;
import org.quartz.TriggerBuilder;
import org.quartz.TriggerKey;
import org.springframework.beans.factory.annotation.Autowired;
import org.springframework.scheduling.quartz.SchedulerFactoryBean;
import org.springframework.stereotype.Service;
import com.zfx.task.bean.TaskBean;

@Service
public class TaskService {

    private Scheduler scheduler;

    public TaskService(@Autowired SchedulerFactoryBean schedulerFactoryBean) {
        scheduler = schedulerFactoryBean.getScheduler();
    }

    /**创建和启动定时任务*/
    public void scheduleJob(TaskBean taskBean) throws SchedulerException {
        JobKey jobKey = taskBean.getJobKey();
        JobDataMap jobDataMap = getJobDataMap(taskBean.getJobDataMap());
        String description = taskBean.getDescription();
        Class<? extends Job> jobClass = taskBean.getJobClass();
        String cron = taskBean.getCronExpression();
        JobDetail jobDetail = getJobDetail(jobKey, description, jobDataMap, jobClass);
        Trigger trigger = getTrigger(jobKey, description, jobDataMap, cron);
        scheduler.scheduleJob(jobDetail, trigger);
    }

    /**暂停Job*/
    public void pauseJob(JobKey jobKey) throws SchedulerException {
        scheduler.pauseJob(jobKey);
    }

    /**恢复Job*/
    public void resumeJob(JobKey jobKey) throws SchedulerException {
        scheduler.resumeJob(jobKey);
    }

    /**删除Job*/
    public void deleteJob(JobKey jobKey) throws SchedulerException {
        scheduler.deleteJob(jobKey);
    }

    /**修改Job的cron表达式*/
    public boolean modifyJobCron(TaskBean taskBean) {
        String cronExpression = taskBean.getCronExpression();
```

```java
            if (!CronExpression.isValidExpression(cronExpression)) return false;
            JobKey jobKey = taskBean.getJobKey();
            TriggerKey triggerKey = new TriggerKey(jobKey.getName(), jobKey.getGroup());
            try {
                CronTrigger cronTrigger = (CronTrigger) scheduler.getTrigger(triggerKey);
                JobDataMap jobDataMap = getJobDataMap(taskBean.getJobDataMap());
                if (!cronTrigger.getCronExpression().equalsIgnoreCase(cronExpression)) {
                    CronTrigger trigger = TriggerBuilder.newTrigger()
                            .withIdentity(triggerKey)
                            .withSchedule(CronScheduleBuilder.cronSchedule(cronExpression))
                            .usingJobData(jobDataMap)
                            .build();
                    scheduler.rescheduleJob(triggerKey, trigger);
                }
            } catch (SchedulerException e) {
             e.printStackTrace();
                return false;
            }
            return true;
        }

        /**获取定时任务的定义*/
        public JobDetail getJobDetail(JobKey jobKey, String description, JobDataMap jobDataMap, Class<? extends Job> jobClass) {
            return JobBuilder.newJob(jobClass)
                    .withIdentity(jobKey)
                    .withDescription(description)
                    .setJobData(jobDataMap)
                    .usingJobData(jobDataMap)
                    .requestRecovery()
                    .storeDurably()
                    .build();
        }

        /**获取 Trigger (Job 的触发器,执行规则)*/
        public Trigger getTrigger(JobKey jobKey, String description, JobDataMap jobDataMap, String cronExpression) {
            return TriggerBuilder.newTrigger()
                    .withIdentity(jobKey.getName(), jobKey.getGroup())
                    .withDescription(description)
                    .withSchedule(CronScheduleBuilder.cronSchedule(cronExpression))
                    .usingJobData(jobDataMap)
                    .build();
        }

        public JobDataMap getJobDataMap(Map<?, ?> map) {
            return map == null ? new JobDataMap() : new JobDataMap(map);
        }
    }
```

Scheduler 的生命期,从 SchedulerFactory 创建它时开始,到 Scheduler 调用 shutdown() 方法时结束;Scheduler 被创建后,可以增加、删除、列举 Job 和 Trigger,以及执行其他与调度相关的操作(如暂停 Trigger)。但是 Scheduler 只有在调用 start()方法后,才会真正地触发 Trigger,即执行 Job。Quartz API 的关键接口及释义如表 10-5 所示。

表 10-5 Quartz API 的关键接口及释义

接　口	释　义
Scheduler	与调度程序交互的主要 API
Job	希望由调度程序执行的组件实现的接口
JobDetail	用于定义作业的实例
Trigger	触发器,定义执行给定作业的计划的组件
JobBuilder	用于定义/构建 JobDetail 实例
TriggerBuilder	用于定义/构建触发器实例

在创建 JobDetail 时,将要执行的 Job 的类名传给 JobDetail,所以 Scheduler 知道要执行何种类型的 Job;每次当 Scheduler 执行 Job 时,在调用其 execute(…)方法前会创建该类的一个新实例;执行完毕,对该实例的引用就被丢弃了,实例会被作为垃圾回收。该执行策略带来的一个后果是,Job 必须有一个无参的构造函数(当使用默认的 JobFactory 时);另一个后果是,在 Job 类中不应该定义有状态的数据属性,因为在 Job 的多次执行过程中,属性的值可能不会被保留。

JobDataMap.class:JobDataMap 中可以包含不限量的(序列化的)数据对象,在 Job 实例执行时,可以使用其中的数据;JobDataMap 是 Java Map 接口的一个实现,额外增加了一些便于存取基本类型数据的方法。

10.4.8 增加控制层

为方便测试,增加 TaskController.java 接口类,接口包括执行 Quartz 的启动、暂停、恢复、删除、修改 cron 表达式等相关操作,其代码如下。

```
package com.zfx.controller;
import org.quartz.JobKey;
import org.quartz.SchedulerException;
import org.springframework.beans.factory.annotation.Autowired;
import org.springframework.web.bind.annotation.GetMapping;
import org.springframework.web.bind.annotation.RequestMapping;
import org.springframework.web.bind.annotation.RestController;
import com.zfx.task.TaskJob;
import com.zfx.task.bean.TaskBean;
import com.zfx.task.service.TaskService;

@RestController
public class TaskController {
    @Autowired
```

```java
    private TaskService taskService;

    //任意设置 Key
    private final JobKey jobKey = JobKey.jobKey("oneJob", "Group1");

    /**启动*/
    @RequestMapping("/startJob")
    public String startHelloWorldJob() throws SchedulerException {
        TaskBean task = new TaskBean();
        task.setJobKey(jobKey);;
        task.setCronExpression("0/2 * * * * ? ");
        task.setJobClass(TaskJob.class);
        task.setDescription("startJob");
        taskService.scheduleJob(task);
        return "启动任务调度成功";
    }

    /**暂停*/
    @GetMapping("/pauseJob")
    public String pauseHelloWorldJob() throws SchedulerException {
        taskService.pauseJob(jobKey);
        return "暂停任务调度成功";
    }

    /**恢复*/
    @GetMapping("/resumeJob")
    public String resumeHelloWorldJob() throws SchedulerException {
        taskService.resumeJob(jobKey);
        return "恢复任务调度成功";
    }

    /**删除*/
    @GetMapping("/deleteJob")
    public String deleteHelloWorldJob() throws SchedulerException {
        taskService.deleteJob(jobKey);
        return "删除任务调度成功";
    }

    /**修改 hello world 的 cron 表达式*/
    @GetMapping("/modifyJobCron")
    public String modifyHelloWorldJobCron() {
        TaskBean task = new TaskBean();
        task.setJobKey(jobKey);;
        task.setCronExpression("0/5 * * * * ? ");
        task.setJobClass(TaskJob.class);
        task.setDescription("这是一个测试的 任务");
        try {
            taskService.scheduleJob(task);
```

```
        } catch (SchedulerException e) {
            e.printStackTrace();
        }
        if (taskService.modifyJobCron(task))
            return "任务调度修改成功";
        else
            return "任务调度修改失败";
    }
}
```

10.4.9 当前项目结构

当前项目结构如图 10-5 所示。Spring Boot 微服务整合 Quartz 分布式，只需要更改 TaskBean.class、TaskService.class、TaskJob.class、application.yml 四个文件即可，同时为了测试效果，复制图 10-5 中的 boot_async_00 项目并粘贴，得到完全相同的 boot_async_00-2 项目，如图 10-6 所示，并修改图 10-6 中 boot_async_00-2 项目的端口号，以免发生端口号冲突。

图 10-5

图 10-6

10.4.10 运行效果

图 10-7 所示的 boot_async_00 项目在启动后自行执行了相关任务调度，而图 10-8 的 boot_async_00-2 项目虽然也被启动了，但在只修改端口号，代码完全相同的情况下，图 10-8

中的 boot_async_00-2 项目没有执行任务调度函数。在多台服务器同时含有某一任务调度的情况下，只有一台机器正常进行任务调度，说明此时任务调度的分布式已经执行成功。

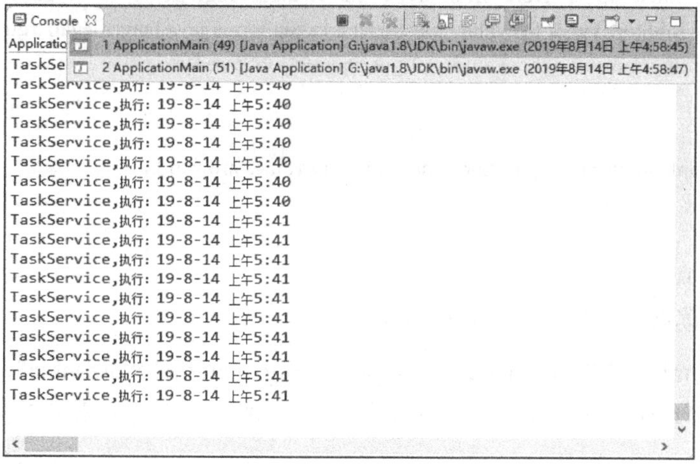

图 10-7

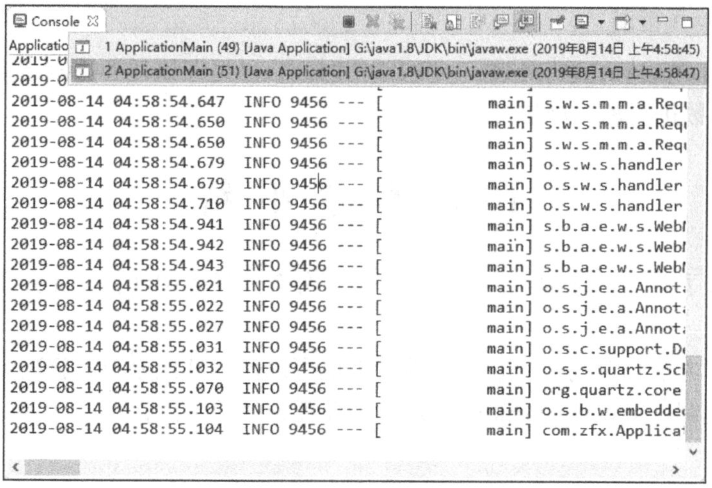

图 10-8

10.4.11 实例易错点

1. Quartz 读取数据库方式异常

若启动程序后报下述异常，则需检查 application.yml 文件中编写的 spring.quartz.job-store-type 是否为 JDBC。

```
Description:
Failed to bind properties under 'spring.quartz.job-store-type' to org.springframework.boot.autoconfigure.quartz.JobStoreType:

    Property: spring.quartz.job-store-type
```

```
Value: jpa
Origin: class path resource [application.yml]:30:21
Reason: Failed to convert from type [java.lang.String] to type [org.springframework.boot.
autoconfigure.quartz.JobStoreType] for value 'jpa'; nested exception is java.lang.IllegalArgumentException: No
enum constant org.springframework.boot.autoconfigure.quartz.JobStoreType.jpa


Action:
Update your application's configuration. The following values are valid:

JDBC
MEMORY
```

2．Quartz 读取数据库无法找到相应表

application.yml 文件中 spring.quartz.properties.org.quartz.jobStore.tablePrefix 字段存储的是，Quartz 读取 MySQL 内管理 Quartz 分布式表的表前缀，此时应查看是否表前缀读取错误。

```
org.quartz.impl.jdbcjobstore.LockException: Failure obtaining db row lock: Table 'justLearn.QRTZ_
22LOCKS' doesn't exist
```

3．ShutDown 错误

Quartz 的 ShutDown()函数需要一定的预留时间，以保证该任务正常结束，此时可调用 Thread.sleep()函数进行辅助操作。

10.5 本章小结

10.1 节整合了单点任务调度，10.4 节整合了分布式任务调度，以此介绍了任务调度的分布式方法。

Quartz 使用方法十分简捷，在实际项目中也可以使用其他分布式解决方案的框架。其实更重要的是理解任务调度的分布式原理，而非死记硬背 API。

10.6 习　题

1．创建 Spring Boot 工程，整合@Scheduled 任务调度。

2．创建 Spring Boot 工程 1 与 Spring Boot 工程 2，使两个工程同时整合 Quartz 达到分布式任务调度的效果。

第 11 章 微服务的文件上传与分布式文件管理

文件上传是一套应用程序开发的基本功能，本章将通过 Spring Boot 实现文件上传功能，从本地将文件上传至指定位置。本章还将通过 FastDFS 方案将文件分离到单独服务器进行统一管理，达到分布式文件管理的目的。

11.1 文件上传/下载原理

文件上传是指将文件转化成二进制数据，通过 HTTP 请求将二进制数据传输到服务器中，并在服务器中创建相应空间，然后将二进制文件写入，转换成文件。在前端发送 Request 请求给 Java 后台时，会在 Request Headers 消息头中增加相应参数。Request Headers 常用的参数及释义如表 11-1 所示。

表 11-1 Request Headers 常用的参数及释义

参　数	释　义
POST /index HTTP/1.1	请求方式、文件名、HTTP 版本号
Host: localhost:8080	请求地址
Connection: keep-alive	Connection 决定当前的事务完成后，是否关闭网络连接。如果该值为"keep-alive"，网络连接就是持久的且不会关闭，使得对同一个服务器的请求可以继续在该连接上完成
Content-Length: 557	发送给 HTTP 服务器的长度
Origin: http://localhost:8080	起源，即来自哪里
X-Requested-With: XMLHttpRequest	表明是 Ajax 异步请求
Accept-Encoding: gzip, deflate, br	浏览器申明自己接收的编码方式，通常指定压缩、是否支持压缩、支持压缩的方式（gzip/default）
Accept-Language: zh-CN,zh;q=0.9	浏览器申明自己接收的语言
Accept:application/json,text/javascript, */*; q=0.01	浏览器接收的媒体类型。application/json, 代表接收 json 类型数据；text/javascript */* 代表浏览器可以处理所有类型
Content-Type: application/json; charset=UTF-8	浏览器接收的内容类型、字符集

表 11-1 中的 Content-Type: application/json 参数代表浏览器接收 json 类型的字符串数据，在前台可以直接使用 Ajax 对 json 格式数据进行解析。而除 application/json 外，Content-Type 常用的参数及释义如表 11-2 所示。

表 11-2 Content-Type 常用的参数及释义

参　　数	释　　义
text/html	html 格式
text/plain	纯文本格式
text/xml	xml 格式
image/gif	gif 图片格式
image/jpeg	jpg 图片格式
image/png	png 图片格式
application/xml	xml 数据格式
application/json	json 数据格式
application/pdf	pdf 格式
application/msword	Word 文档格式
application/octet-stream	二进制流数据（如常见的文件下载）
multipart/form-data	文件上传

表 11-2 中的 multipart/form-data 参数即是常见的数据上传提交方式，在使用表单提交时必须使 method 提交方式为 post，encType=multipart/form-data。encType 属性管理表单，报文如下。

```
POST http://localhost:8080/index HTTP/1.1
Content-Type:multipart/form-data
Content-Disposition: form-data; name="text"
Content-Disposition: form-data; name="file"; filename="chrome.png"
Content-Type: image/png
```

form 表单的 enctype 属性规定发送到服务器前应该如何对表单数据进行编码（content-type），并且告知服务器请求正文的 MIME 类型，encType 属性一共含有 3 种参数，如表 11-3 所示。

表 11-3 form 表单的 encType 属性参数及释义

参　　数	释　　义
application/x-www-form-urlencoded	在发送前编码所有字符（默认），只处理表单域中的 Value 属性值，采用该编码方式的表单会将表单域的值处理成 URL 编码方式
multipart/form-data	不对字符编码，每个表单项分割为一个部件，表单以二进制流的方法处理表单数据，该编码方式会将文件域指定文件的内容也封装到请求参数里
text/plain	空格转换为 "+" 加号，但不对特殊字符编码，主要适用直接通过表单发送邮件的方式

在 Java 项目文件上传，前端请求后台程序时，首先依靠将消息头转换成彼此认知的方式，告知后台系统该请求是文件上传请求，流程如下。

（1）表单提交。

（2）将文件转换成二进制。

（3）用 Servlet 接收 Request 请求，通过 Request Header 消息头认知该请求是文件上传请求，并接收相应的二进制文件。

Java 应用程序文件上传，继承并重写 Servlet 代码如下。

```
protected void doPost(HttpServletRequest request, HttpServletResponse response)
    throws ServletException, IOException {
        try {
            // 配置上传参数
            DiskFileItemFactory factory = new DiskFileItemFactory();
            ServletFileUpload upload = new ServletFileUpload(factory);
            // 解析请求的内容提取文件数据
            List<FileItem> formItems = upload.parseRequest(request);
            // 迭代表单数据
            for (FileItem item : formItems) {
                // 处理不在表单中的字段
                if (!item.isFormField()) {
                    String fileName = item.getName();
                    //定义上传文件的存放路径
                    String path = request.getServletContext().getRealPath("/uploadFiles");
                    //定义上传文件的完整路径
                    String filePath = String.format("%s/%s",path,fileName);
                    File storeFile = new File(filePath);
                    // 在控制台输出文件的上传路径
                    System.out.println(filePath);
                    // 保存文件到硬盘
                    item.write(storeFile);
                }
            }
        } catch (Exception e) {
        }
    }
```

11.1.1 SpringMVC 文件上传原理

Multipart 格式的数据会将一个表单拆成多个部分（part），每个部分对应一个输入区域，在一般表单输入区域中，Multipart 对应的部分会放置文本类型的数据，但是如果上传文件，Multipart 对应的部分是二进制的。正常 SpringMVC 在接收到 Request 请求时，响应步骤如下。

（1）用户发送请求至前端控制器 DispatcherServlet。

（2）DispatcherServlet 接收到请求，调用 HandlerMapping 处理器映射器。

（3）处理器映射器找到具体的处理器（可根据 XML 配置、注解进行查找），生成处理器对象及处理器拦截器（若有则生成），一并返回给 DispatcherServlet。

（4）DispatcherServlet 调用 HandlcrAdapter 处理器适配器。

（5）HandlerAdapter 经过适配调用具体的处理器（Controller，称为后端控制器）。

（6）Controller 执行完成返回 ModelAndView。

（7）HandlerAdapter 将 Controller 执行结果 ModelAndView 返回给 DispatcherServlet。

（8）DispatcherServlet 将 ModelAndView 传给 ViewReslover 视图解析器。

（9）ViewReslover 解析后返回具体 View。

（10）DispatcherServlet 根据具体 View 渲染视图（即将模型数据填充到视图中）。

（11）DispatcherServlet 响应用户。

当收到 Multipart 文件上传请求时，DispatcherServlet 的 checkMultipart()方法会调用 MultipartResolver 的 isMultipart()方法，判断请求数据中是否包含文件。若请求数据中包含文件，则调用 MultipartResolver 的 resolveMultipart()方法对请求数据进行解析，然后将文件数据解析成 MultipartFile 并封装在 MultipartHttpServletRequest（MultipartHttpServletRequest 继承了 HttpServletRequest）对象中，SpringMVC 通过 MultipartResolver 策略接口告知 DispatcherServlet 控制器，应如何解析该 Multipart 请求，最后返回响应。SpringMVC 上传代码如下。

```html
<!--表单-->
<form action="upload" enctype="multipart/form-data" method="post">
        <table>
            <tr>
                <td>请选择文件:</td>
                <td><input type="file" name="file"></td>
            </tr>
            <tr>
                <td><input type="submit" value="上传"></td>
            </tr>
        </table>
</form>

//Controller 类
@RequestMapping(value="/upload",method=RequestMethod.POST)
public String upload(HttpServletRequest request,
        @RequestParam("file") MultipartFile file) throws Exception {
    try {
            //获取输出流
            OutputStream os=null;
            os=new FileOutputStream("E:/"+file.getOriginalFilename());
            //获取输入流 CommonsMultipartFile 中可以直接得到文件的流
            InputStream is=file.getInputStream();
            int temp;
            //一个一个字节读取并写入
            while((temp=is.read())!=(-1)){
                os.write(temp);
            }
            os.flush();
```

```
            os.close();
            is.close();
        } catch (FileNotFoundException e) {
            e.printStackTrace();
        }
        return "/success";
    }
```

有时可能感觉整个程序运转速度较慢，这是因为应用程序在接收完整的 Request Headers 数据后，需要上传全部的二进制文件（MultipartFile）后才会对其进行解析。

11.1.2 文件下载原理

文件下载的本质是将文件变成 IO 流，通过 Response 响应将数据返回给前台。后台使用 HttpServletResponse.setContentType 方法设置 Content-Type 字段的值，通常设置为 application/octet-stream 或 application/x-msdownload，决定客户端服务器以何种方式接收返回的信息。使用 HttpServletResponse.setHeader 方法设置 Content-Disposition 头的值为 "attachment;filename=文件名"，浏览器通过附件的形式获取到用户上传的文件。在读取下载文件时，使用 HttpServletResponse.getOutputStream 方法返回 ServletOutputStream 对象，向客户端写入附件文件的内容。Java 应用程序文件下载代码如下。

```
public HttpServletResponse download(String path, HttpServletResponse response) {
    try {
        // path 是指欲下载的文件的路径
        File file = new File(path);
        // 取得文件名
        String filename = file.getName();
        // 取得文件的后缀名
        String ext = filename.substring(filename.lastIndexOf(".") + 1).toUpperCase();
        // 以流的形式下载文件
        InputStream fis = new BufferedInputStream(new FileInputStream(path));
        byte[] buffer = new byte[fis.available()];
        fis.read(buffer);
        fis.close();
        // 清空 response
        response.reset();
        // 设置 response 的 Header
        response.addHeader("Content-Disposition","attachment;filename="+new String(filename.getBytes()));
        response.addHeader("Content-Length", "" + file.length());
        OutputStream toClient = null;
        toClient = new BufferedOutputStream(response.getOutputStream());
        response.setContentType("application/octet-stream");
        toClient.write(buffer);
        toClient.flush();
        toClient.close();
    } catch (IOException ex) {
```

```
            ex.printStackTrace();
        }
    return response;
}
```

若文件更小,也可以使用直接配置映射路径的方式,给前端展示并直接下载。

11.2 【实例】微服务的单文件和多文件上传

11.2.1 实例背景

本实例创建 boot_08_1 项目,用 Spring Boot 整合文件上传,在前台使用 JSP 页面上传文件后,微服务的接口将上传的文件存储到指定位置。编写 pom.xml 文件代码如下。

```xml
<project xmlns="http://maven.apache.org/POM/4.0.0" xmlns:xsi="http://www.w3.org/2001/XMLSchema-instance" xsi:schemaLocation="http://maven.apache.org/POM/4.0.0 http://maven.apache.org/xsd/maven-4.0.0.xsd">
    <modelVersion>4.0.0</modelVersion>
    <groupId>org.zfx.boot</groupId>
    <artifactId>boot_07_1</artifactId>
    <version>0.0.1-SNAPSHOT</version>
    <parent>
        <groupId>org.springframework.boot</groupId>
        <artifactId>spring-boot-starter-parent</artifactId>
        <version>2.0.0.RELEASE</version>
    </parent>
    <dependencies>
        <dependency>
            <groupId>org.springframework.boot</groupId>
            <artifactId>spring-boot-starter-web</artifactId>
        </dependency>
        <!-- servlet 依赖. -->
        <dependency>
            <groupId>javax.servlet</groupId>
            <artifactId>javax.servlet-api</artifactId>
            <scope>provided</scope>
        </dependency>
        <dependency>
            <groupId>javax.servlet</groupId>
            <artifactId>jstl</artifactId>
        </dependency>
        <!-- tomcat 的支持. -->
        <dependency>
            <groupId>org.springframework.boot</groupId>
            <artifactId>spring-boot-starter-tomcat</artifactId>
```

```xml
            <scope>provided</scope>
        </dependency>
        <dependency>
            <groupId>org.apache.tomcat.embed</groupId>
            <artifactId>tomcat-embed-jasper</artifactId>
            <scope>provided</scope>
        </dependency>
        <!-- 添加热部署，开发者模式 -->
        <dependency>
            < groupId>org.springframework.boot</groupId>
            <artifactId>spring-boot-devtools</artifactId>
            <optional>true</optional>
        </dependency>
    </dependencies>
    <build>
        <plugins>
            <plugin>
                <artifactId>maven-compiler-plugin</artifactId>
                <configuration>
                    <source>1.8</source>
                    <target>1.8</target>
                </configuration>
            </plugin>
        </plugins>
    </build>
</project>
```

maven-compiler-plugin 插件：如果没有设置 Maven 项目管理工具使用何种版本的 JDK 进行编译，Maven 就会用 maven-compiler-plugin 默认的 JDK 版本处理（默认为 JDK 1.5），此时易出现版本不匹配等相关问题，可能导致编译不通过。maven-compiler-plugin 插件可以指定项目源码的 JDK 版本、编译后的 JDK 版本及编码。

spring-boot-devtools 为热部署，devtools 会监听所有文件夹下的代码，若发生改变，则会直接重启 Spring Boot 应用程序。

11.2.2　编写 application.properties 资源配置文件

在 application.properties 资源配置文件中设置 Spring Boot 应用程序，关于 JSP 的相关配置代码如下。

```
#上传文件总的最大值
spring.servlet.multipart.max-request-size=10MB
#单个文件的最大值
spring.servlet.multipart.max-file-size=10MB
## jsp 页面位置前缀后缀
spring.mvc.view.prefix=/WEB-INF/jsp/
```

```
spring.mvc.view.suffix=.jsp
#关闭默认模板引擎
spring.thymeleaf.cache=false
spring.thymeleaf.enabled=false
```

即使不关闭 thymeleaf 模板引擎，也不会出现太大的问题。其实 Spring Boot 本身并不推荐使用 JSP，而推荐使用 thymeleaf、freemaker 引擎模板等，所以对其关闭可避免出现不必要的错误。

11.2.3 编写相关接口

为方便测试，编写 IndexController.java 接口类。IndexController.java 接口类分别提供接收文件上传与多文件上传接口，其代码如下。

```java
package org.zfx.controller;
import java.io.File;
import java.io.IOException;
import java.util.List;
import java.util.TreeMap;
import javax.servlet.http.HttpServletRequest;
import org.springframework.stereotype.Controller;
import org.springframework.web.bind.annotation.RequestMapping;
import org.springframework.web.bind.annotation.RequestParam;
import org.springframework.web.bind.annotation.ResponseBody;
import org.springframework.web.multipart.MultipartFile;
import org.springframework.web.multipart.MultipartHttpServletRequest;

@Controller
public class IndexController {

    /**
     * 单文件上传
     */
    @RequestMapping("/uploadOne")
    @ResponseBody
    public TreeMap<String, Object> uploadOne(@RequestParam("file") MultipartFile file) {
        TreeMap<String, Object> resultMap = new TreeMap<>();

        if (file.isEmpty()) {
            resultMap.put("success", "上传失败，请选择文件");
            return resultMap;
        }
        String fileName = file.getOriginalFilename();
        String filePath = "F:\\book\\2\\";
        File dest = new File(filePath + fileName);
        try {
            file.transferTo(dest);
```

```java
            resultMap.put("success", "上传成功");
            return resultMap;
        } catch (IOException e) {
            resultMap.put("success", "上传失败,IOException");
            return resultMap;
        }
    }

    /**
     * 多文件上传
     */
    @RequestMapping("/uploadMany")
    @ResponseBody
    public TreeMap<String, Object> uploadOne(HttpServletRequest request) {
        List<MultipartFile> files = ((MultipartHttpServletRequest) request).getFiles("file");
        TreeMap<String, Object> resultMap = new TreeMap<>();

        String filePath = "F:\\book\\2\\";
        for (int i = 0; i < files.size(); i++) {
            MultipartFile file = files.get(i);
            if (file.isEmpty()) {
                resultMap.put("success" + ++i,"第"+ i + "个文件为空");--i;
            }

            String fileName = file.getOriginalFilename();
            File dest = new File(filePath + fileName);
            try {
                file.transferTo(dest);
                resultMap.put("success" + ++i,"第"+ i + "个文件上传成功");--i;
            } catch (IOException e) {
                resultMap.put("success" + ++i,"第"+ i + "个文件上传失败");--i;
            }
        }
        return resultMap;
    }

    @RequestMapping({"/index","/"})
    public String index() {
        return "uploadpage";
    }
}
```

MultipartFile 接口:org.springframework.web.multipart 接口继承于 org.springframework.core.ioInputStreamSource 流对象接口,定义了与文件存储相关的内容,如文件内容存储在硬盘上还是内存上等。MultipartFile 接口的函数及释义如表 11-4 所示。

表 11-4 MultipartFile 接口的函数及释义

函　　数	释　　义
String getName()	返回文件的名称
String getOriginalFilename()	返回客户端文件系统中的原始文件名，getOriginalFilename 可能包含路径信息，具体取决于浏览器；若未选择文件，则返回空字符串
String getContentType()	返回文件的内容类型
Boolean isEmpty()	判断是否为空
Long getSize()	返回文件大小
byte[] getBytes()	返回文件的字节数组，若为空，则返回空的字节数组
InputStream getInputStream()	获得文件的输入流
Void transferTo()	将接收到的文件传输到给定的目标文件；transferTo 可以移动文件系统中的文件，将文件复制到文件系统；如果目标文件已存在，将先将其删除

MultipartFile 接口的实现类及释义如表 11-5 所示。

表 11-5 MultipartFile 接口的实现类及释义

实　现　类	释　　义
org.springframework.web.multipart.commons.CommonsMultipartFile	Apache Commons 文件上载的实现
org.springframework.web.multipart.support.StandardMultipartHttpServletRequest$StandardMultipartFile	Spring 多部分文件适配器，包装 Servlet 3.0 部件对象

11.2.4 编写前台页面

为方便测试，编写前台的 src/main/webapp/WEB-INF/jps/uploadpage.jsp 页面，对接 IndexController.java 类中的文件上传与多文件上传接口，代码如下。

```
<%@ page contentType="text/html;charset=UTF-8" pageEncoding="UTF-8" %>
<!DOCTYPE html>
<html>
<head>
    <meta http-equiv="Content-type" content="text/html; charset=UTF-8">
    <title>文件上传</title>
</head>
<body>
    <form method="post" action="/uploadOne" enctype="multipart/form-data">
        <input type="file" name="file"><br>
        <input type="submit" value="提交一个文件表单">
    </form>

    <hr/>
    <form method="post" action="/uploadMany" enctype="multipart/form-data">
```

```
            <input type="file" name="file"><br>
            <input type="file" name="file"><br>
            <input type="file" name="file"><br>
            <input type="file" name="file"><br>
            <input type="submit" value="提交多个文件表单">
        </form>
    </body>
</html>
```

<form>表单：<form>标签用于为用户输入创建 HTML 表单。表单包含 input 元素，如文本字段、复选框、单选框、提交按钮等。表单还可以包含 menus、textarea、fieldset、legend 和 label 元素。表单用于向服务器传输数据。HTML5 的<form>标签常用的参数及释义如表 11-6 所示。

表 11-6 HTML5 的<form>标签常用的参数及释义

参　数	值	释　义
accept-charset	charset_list	规定服务器可处理的表单数据字符集
action	URL	规定当提交表单时向何处发送表单数据
autocomplete	on off	规定是否启用表单的自动完成功能
enctype	application/x-www-form-urlencoded multipart/form-data text/plain	规定在发送表单数据前如何对其进行编码
method	get post	规定用于发送 form-data 的 HTTP 方法
name	form_name	规定表单的名称
novalidate	novalidate	若使用该属性，则提交表单时不进行验证
target	_blank _self _parent _top framename	规定在何处打开 action URL

11.2.5 当前项目结构

如图 11-1 所示，本项目将 jps 页面放置在 src/main/webapp/WEB-INF/jps 目录下，其中的文件夹都需要自行创建，才可在微服务中正常显示。

Spring Boot 微服务默认提供静态文件存储地址如下。

（1）/boot_08_1/src/main/resources 目录下。

（2）/boot_08_1/src/main/resources/static 目录下。

（3）/boot_08_1/src/main/resources/public 目录下。

（4）/boot_08_1/src/main/resources/META-INF 目录下。

（5）/boot_08_1/src/main/webapp 目录下。

Spring Boot 微服务默认提供静态文件存储地址，如图 11-2 所示。

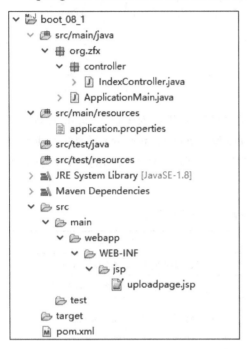

图 11-1

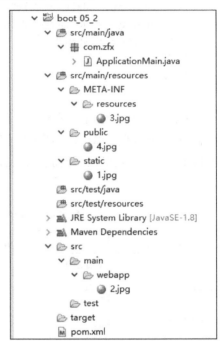

图 11-2

Spring Boot 微服务推荐将静态文件，如 Freemaker、JSP、CSS、JavaScript、HTML 等，放置在 resources 文件夹下。URL 可以直接对其进行调取。

11.2.6　运行结果

（1）打开展示页面，页面地址为 localhost:8080/index，效果如图 11-3 所示。

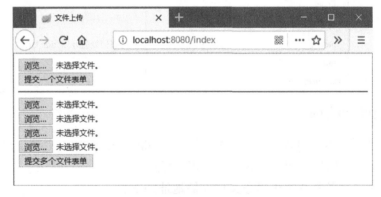

图 11-3

（2）选择单文件上传并执行，效果如图 11-4 所示。

（3）选择多文件上传，并上传全部文件成功，效果如图 11-5 所示。

图 11-4

图 11-5

（4）选择多文件上传，只上传第 2 个和第 4 个文件，效果如图 11-6 所示。

图 11-6

在上传时可看到本地磁盘内已被写入了相关文件。

11.2.7 实例易错点

1. 微服务永不能编写 error.jsp 页面

在集成 JSP 时，切忌使用 error.jsp 作为页面处理，因为 Spring Boot 微服务本身含有一个渲染 error 的页面，一旦返回 error 时，会默认为返回错误的渲染页面 error，error 页面需要使用 timestamp 时间戳，没有传参即会报错。SpringMVC 的 error 渲染页面代码全路径地址如下。

org.springframework.boot.autoconfigure.web.ErrorMvcAutoConfiguration.SpelView

渲染页面部分代码如图 11-7 所示。

```
@Bean
public static PreserveErrorControllerTargetClassPostProcessor preserveErrorControllerTargetClassPostProcessor() {
    return new PreserveErrorControllerTargetClassPostProcessor();
}

@Configuration
@ConditionalOnProperty(prefix = "server.error.whitelabel", name = "enabled", matchIfMissing = true)
@Conditional(ErrorTemplateMissingCondition.class)
protected static class WhitelabelErrorViewConfiguration {

    private final SpelView defaultErrorView = new SpelView(
            "<html><body><h1>Whitelabel Error Page</h1>"
                    + "<p>This application has no explicit mapping for /error, so you are seeing this as a fallback.</p>"
                    + "<div id='created'>${timestamp}</div>"
                    + "<div>There was an unexpected error (type=${error}, status=${status}).</div>"
                    + "<div>${message}</div></body></html>");

    @Bean(name = "error")
    @ConditionalOnMissingBean(name = "error")
    public View defaultErrorView() {
        return this.defaultErrorView;
    }
}
```

图 11-7

本实例使用 SpringMVC 下的 Modelandview 视图控制类，也可采用其他相关的 Request、Map、String 等方式。

2．微服务项目名设置

在 Spring Boot 微服务的 application 资源配置文件中需增加相关配置。

如果 starter 是 2.x.x 版本，增加配置如下。

server.servlet.context-path=/zfx

如果 starter 是 1.x.x 版本，增加配置如下。

server.context-path=/zfx

若版本与配置不对应，则无法增加微服务的项目名。设置微服务的项目名后执行的效果如图 11-8 与图 11-9 所示。

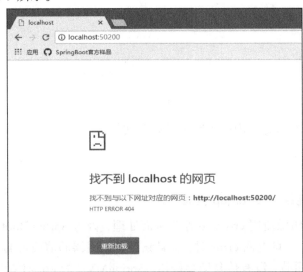

图 11-8

图 11-9

11.3 分布式文件管理

11.3.1 分布式文件管理特性

分布式文件管理、分布式文件系统、分布式存储系统都是指将数据分散存储在多台独立服务器上。传统的文件管理只存储在自身的服务器上。分布式文件管理会单独设置一个服务器作为文件存储服务器（又称为文件管理服务器），所有的文件资源、文件都会存储在文件存储服务器上。

11.3.2 分布式文件管理解决的问题

分布式文件管理解决了单节点存储上线的难度，即如果存储空间不足，只用手工方式删除或增加服务器十分困难，而使用分布式文件管理后只需要增加服务器数量即可，文件系统会为用户直接进行分散存储，保证了数据量的扩展性。

分布式文件管理使用多台服务器共同存储，随着服务器数量的增加，服务器出现的概率在逐步增加。通常，分布式文件管理会将一份数据复制多份，并分别存储在不同的服务器上，一旦某个存储系统或存储服务器出现了异常，都不会影响整体的分布式文件管理，保证了服务器的高可用性。

分布式文件管理的解决方案通常会配合事务，存储各文件的修改时间、文件大小等，保证多台服务器下各文件的一致性。

11.3.3 分布式文件管理解决方案

分布式文件管理的解决方案有很多，也涉及大数据的部分领域，如大数据分布式文件系统 Hadoop HDFS、OpenStack 的对象存储 Swift、Facebook 用于图片存储的 Haystack、Facebook 的暖性 BLOB 存储、OpenStack 块存储 Cinder、Lustre、Gluster、用户空间文件系统 FUSE、Ceph、通用并行文件系统 GPFS、全局联合文件系统 Global GFFS。

11.4 FastDFS 解决方案

FastDFS 是开源的轻量级分布式文件系统，它可对文件进行管理，包括文件存储、文件同步、文件访问（文件上传、文件下载）等，解决了大容量存储和负载均衡的问题，特别适合以文件为载体的在线服务，如相册网站、视频网站等。

FastDFS 为互联网量身定制，充分考虑了冗余备份、负载均衡、线性扩容等机制，并注重高可用、高性能等指标，使用 FastDFS 很容易搭建一套高性能的文件服务器集群，提供文件上传、下载等服务。

FastDFS 系统有三个角色：跟踪服务器（Tracker Server）、存储服务器（Storage Server）和客户端（Client），如表 11-7 所示。

表 11-7 FastDFS 系统角色及作用

角色	作用
Tracker Server	跟踪服务器，主要做调度工作，起到均衡的作用；负责管理所有的 Storage Server 和 Group，每个 Storage 在启动后会连接 Tracker，告知自己所属 Group 等信息，并保持周期性心跳
Storage Server	存储服务器，主要提供容量和备份服务；以 Group 为单位，每个 Group 内可有多台 Storage Server，数据互为备份
Client	客户端，上传或下载数据的服务器，也是项目所部署的服务器

11.4.1 FastDFS 的存储策略

为了支持大容量，存储节点（服务器）采用了分卷（或分组）的组织方式。存储系统由一个或多个卷组成，卷与卷之间的文件是相互独立的，所有卷的文件容量进行累加就是整个存储系统中的文件容量。一个卷可由一台或多台存储服务器组成，一个卷下的存储服务器中的文件都是相同的，卷中的多台存储服务器起到了冗余备份和负载均衡的作用。

在卷中增加服务器时，同步已有的文件由系统自动完成，同步完成后，系统自动将新增服务器切换到线上提供服务。当存储空间不足或即将耗尽时，可以动态添加卷，只需要增加一台或多台服务器，并将它们配置为一个新的卷就扩大了存储系统的容量。

11.4.2 FastDFS 的文件上传过程

FastDFS 向使用者提供基本文件访问接口，如 upload、download、append、delete 等，以客户端库的方式提供给用户使用。

Storage Server 会定期向 Tracker Server 发送自己的存储信息。当 Tracker Server Cluster 中有多个 Tracker Server 时，各 Tracker 之间的关系是对等的，所以客户端上传时可以选择任意一个 Tracker。

当 Tracker 收到客户端上传文件的请求时，会为该文件分配一个可以存储文件的 Group，当选定 Group 后，要决定给客户端分配 Group 中的哪一个 Storage Server。当分配好 Storage Server 后，客户端向 Storage 发送写文件请求，Storage 将先为文件分配一个数据存储目录，然后为文件分配一个编号，最后根据以上信息生成文件名存储文件。

11.4.3 FastDFS 的文件同步过程

写文件时，客户端将文件写到 Group 内的一个 Storage Server 中即认为写文件成功。Storage Server 写完文件后，会由后台线程将文件同步到同一个 Group 内中其他的 Storage Server 中。

每个 Storage 写完文件后，同时会写一份 Binlog，Binlog 中不包含文件数据，只包含文件名等元信息，Binlog 用于后台同步，Storage 会记录向 Group 内其他 Storage 同步的进度，以便重启后能按上次的进度继续同步；进度以时间戳的方式记录，因此最好能保证集群中所有 Server 的时钟保持同步。

Storage 的同步进度会作为元数据的一部分汇报到 Tracker 上，Tracker 在选择读 Storage 时会以同步进度作为参考。

11.4.4　FastDFS 的文件下载过程

客户端上传文件成功后，会得到一个 Storage 生成的文件名，客户端根据该文件名即可访问该文件。与上传文件一样，在下载文件时客户端可以选择任意 Tracker Server。Tracker 发送下载请求给某个 Tracker，必须有文件名信息，Tracker 从文件名中解析文件的 Group、大小、创建时间等信息，然后为该请求选择一个 Storage 用来服务读请求。

11.5　FastDFS 的安装部署

11.5.1　安装 LibFastCommon

LibFastCommon 是一个开源的 C 基础库，也是从 FastDFS 项目中剥离出来的 C 基础库。LibFastCommon 库简单稳定，函数包括字符串、记录器、链、散列、套接字、ini 文件读取器、Base64 编码/解码、URL 编码/解码、快速定时器、Skiplist、对象池等。LibFastCommon 安装如下。

```
## 下载
[root@localhost mysoft]# wget https://github.com/happyfish100/libfastcommon/archive/V1.0.38.tar.gz
## 解压
[root@localhost mysoft]# tar -zxvf V1.0.38
## 进入 libfastcommon 目录
[root@localhost mysoft]# cd libfastcommon-1.0.38/
## 编译
[root@localhost libfastcommon-1.0.38]# ./make.sh
## 安装
[root@localhost libfastcommon-1.0.38]# ./make.sh install
```

通常，libfastcommon.so 安装在/usr/lib64/libfastcommon.so 地址。

11.5.2　安装 FastDFS

使用 wget 命令可从 Github 网站下载 FastDFS 核心项目，其项目依赖 LibFastCommon 与 Nginx，安装 FastDFS 步骤如下。

```
## 下载
[root@localhost mysoft]# wget https://github.com/happyfish100/fastdfs/archive/V5.11.tar.gz
## 解压
```

```
[root@localhost mysoft]# tar -zxvf V5.11
## 进入 fastdfs-5.11 目录
[root@localhost mysoft]# cd fastdfs-5.11/
## 编译
[root@localhost fastdfs-5.11]# ./make.sh
## 安装
[root@localhost fastdfs-5.11]# ./make.sh install
```

安装 FastDFS 核心项目完成后，为其创建 FastDFS 相关路径如下。

```
[root@localhost data]# mkdir -p /data/fastdfs/log
[root@localhost data]# mkdir -p /data/fastdfs/data
[root@localhost data]# mkdir -p /data/fastdfs/tracker
[root@localhost data]# mkdir -p /data/fastdfs/client
```

以下 3 个文件是 FastDFS 示例。

```
/etc/fdfs/client.conf.sample
/etc/fdfs/storage.conf.sample
/etc/fdfs/tracker.conf.sample
```

11.5.3　配置 FastDFS 的跟踪服务器

如表 11-7 所示，FastDFS 的跟踪服务器 Tracker Server 是其三大核心之一。在 /etc/fdfs/tracker.conf.sample 地址有跟踪器的示例文件，而更改 FastDFS 跟踪器的配置文件地址为 /etc/fdfs/tracker.conf，可使用 cp 命令将 /etc/fdfs/tracker.conf.sample 文件复制为 /etc/fdfs/tracker.conf 文件，并更改部分配置如下。

```
## 存储数据和日志文件的基本路径
base_path=/data/fastdfs/tracker
## HTTP 服务端口
http.server_port=9999
## 默认提供服务端口
port=50200
```

http.server_port 的 HTTP 服务端口可设置为 80，但是目前国内对所有 Public Server 服务器禁用了 80 端口，所以此处设置为 9999。如果使用阿里云等云服务器，需在阿里云上配置相应的端口策略。配置 FastDFS 的跟踪器完成后，启动跟踪器，命令如下。

```
## 启动
[root@localhost fdfs]# /usr/bin/fdfs_trackerd /etc/fdfs/tracker.conf start
## 查看状态
[root@localhost fdfs]# ps -ef|grep fdfs
```

如果 Tracker 执行成功，上述查看状态命令输出的信息如下。

```
root        901      1  0 04:42 ?        00:00:00 /usr/bin/fdfs_trackerd /etc/fdfs/tracker.conf start
root        910  36331  0 04:42 pts/0    00:00:00 grep fdfs
```

11.5.4　配置 FastDFS 的数据存储服务器

FastDFS 的数据存储服务器 Storage Server 是其三大核心之一，在/etc/fdfs/storage.conf.sample 地址有数据存储的示例文件，而更改 FastDFS 数据存储的配置文件地址为/etc/fdfs/storage.conf，可使用 cp 命令将/etc/fdfs/storage.conf.sample 文件复制为/etc/fdfs/storage.conf 文件，并更改部分配置如下。

```
## storage 存储 data 和 log 的路径
base_path=/data/fastdfs/data
## 默认组名
group_name=group1
## 默认端口，相同组的 storage 端口号必须一致
port=23000
## 配置一个存储路径
store_path_count=1
store_path0=/data/fastdfs/data
## 配置跟踪器 IP 和端口
tracker_server=192.168.138.157:50200
```

上述配置中 tracker_server 地址应与 11.5.3 节中配置的跟踪器提供服务的 IP 端口相同。配置 FastDFS 的数据存储服务器完成后，启动数据存储 Storage Server，命令如下。

```
## 启动
[root@localhost fdfs]# /usr/bin/fdfs_storaged /etc/fdfs/storage.conf start
## 查看进程
[root@localhost fdfs]# ps -ef|grep fdfs
## 查看启动日志
[root@localhost fdfs]# tail -f /data/fastdfs/data/logs/storaged.log
## 查看 Storage 和 Tracker 是否在通信
[root@localhost fdfs]# /usr/bin/fdfs_monitor /etc/fdfs/storage.conf
```

若成功，启动日志返回如下。

```
[2019-08-18 12:35:47] INFO - file: storage_func.c, line: 257, tracker_client_ip: 192.168.138.157, my_server_id_str: 192.168.138.157, g_server_id_in_filename: -1836406592
[2019-08-18 12:35:48] INFO - file: tracker_client_thread.c, line: 310, successfully connect to tracker server 192.168.138.157:50200, as a tracker client, my ip is 192.168.138.157
[2019-08-18 12:36:18] INFO - file: tracker_client_thread.c, line: 1263, tracker server 192.168.138.157:50200, set tracker leader: 192.168.138.157:50200
```

若成功，查看 Storage Server 和 Tracker Server 是否在通信并返回如下。

```
DEBUG - base_path=/data/fastdfs/data, connect_timeout=30, network_timeout=60, tracker_server_count=1, anti_steal_token=0, anti_steal_secret_key length=0, use_connection_pool=0, g_connection_pool_max_idle_time=3600s, use_storage_id=0, storage server id count: 0
server_count=1, server_index=0
tracker server is 192.168.138.157:50200
```

若成功，查看进程命令输出的信息应如下。

root	901	1	0 04:42 ?	00:00:00 /usr/bin/fdfs_trackerd /etc/fdfs/tracker.conf start
root	1173	1	44 05:01 ?	00:00:03 /usr/bin/fdfs_storaged /etc/fdfs/storage.conf start
root	1176	1138	0 05:01 pts/1	00:00:00 grep fdfs

11.5.5 配置 FastDFS 的客户端并测试

FastDFS 的客户端 Client 是其三大核心之一，在/etc/fdfs/client.conf.sample 地址有客户端的示例文件，而更改 FastDFS 客户端的配置文件地址为/etc/fdfs/client.conf，可使用 cp 命令将/etc/fdfs/client.conf.sample 文件复制为/etc/fdfs/client.conf 文件，并更改部分配置如下。

```
## client 数据和日志目录
base_path=/data/fastdfs/client
## 配置跟踪器 IP 和端口
tracker_server=192.168.138.157:50200
```

上述配置中 tracker_server 地址应与 11.5.3 节中配置的跟踪器提供服务的 IP 端口相同。配置 FastDFS 的客户端完成后，在任意位置上传一张图片（本实例通过 XShell 工具上传 myimg.jpg 图片到/data/img/目录下），通过 FastDFS 客户端工具 fdfs_upload_file（客户端上传工具）将/data/img/myimg.jpg 图片上传到 FastDFS 的存储位置，其上传文件命令如下。

[root@localhost fdfs]# /usr/bin/fdfs_upload_file /etc/fdfs/client.conf /data/img/myimg.jpg

若成功，上述命令的返回信息应如下。

group1/M00/00/00/wKiKnV21zb2AQiv7AABMLJJFkL8028.jpg

此时 FastDFS 的单节点已经搭建完成，接下来依靠 Nginx 搭建 FastDFS 的多节点。FastDFS 的下载也是依靠 Nginx 服务器执行的。

11.5.6 安装 Nginx 部署 FastDFS

Nginx 是一款轻量级的 Web 服务器/反向代理服务器及电子邮件（IMAP/POP3）代理服务器，在 BSD-like 协议下发行。其特点是占用内存少，并发能力强，事实上 Nginx 的并发能力确实比同类型的网页服务器稍强。传统 JavaEE 项目中 Tomcat 的负载均衡使用 Nginx 进行部署，PHP 与 Python Web 等相关项目也多应用 Nginx。

FastDFS 在传统 Nginx 上增加 Fast-Nginx 连接文件，Fast-Nginx 将作为 Nginx 下的一个模块执行，FastDFS 依靠 Fast-Nginx 与 Nginx 连接，并且 fastdfs-nginx 模块可以重定向文件链接到源服务器读取文件，避免客户端由于复制延迟导致的文件无法访问错误。

下载 Nginx 压缩包后，如果 CentOS 和 RedHat 版本过低，可能缺少部分 Nginx 依赖，以下命令可以增加 Nginx 的相关依赖。

```
[root@localhost data]# yum install pcre
[root@localhost data]# yum install pcre-devel
[root@localhost data]# yum install zlib
```

```
[root@localhost data]# yum install zlib-devel
[root@localhost data]# yum install openssl
[root@localhost data]# yum install openssl-devel
```

通常，Nginx 相关命令安装在/usr/local/nginx 地址，然后下载 fastdfs-nginx-module 模块。使用 mv 命令对其重命名，使用 unzip 命令对其解压，解压后将/data/fastdfs-nginx-module/src/mod_fastdfs.conf 复制到/etc/fdfs 目录下，修改/etc/fdfs/mod_fastdfs.conf 配置文件代码如下。

```
## 链接超时
connect_timeout=20
## 配置跟踪器 IP 和端口
tracker_server=192.168.138.157:50200
## 路径包含 group
url_have_group_name = true
# 必须和 Storage 配置相同
store_path0=/data/fastdfs/data
```

修改/etc/fdfs/mod_fastdfs.conf 配置文件后，解压缩 Nginx 压缩包，配置 Nginx 相关内容，修改/data/nginx/conf/nginx.conf 文件。注意，修改/data 目录下的 conf 文件配置，而不是/usr/local/nginx/conf/nginx.conf 配置文件。如果文件修改错误，可能导致 Nginx 的 worker 进程无法正确启动，nginx.conf 配置文件修改部分代码如下。

```
server {
listen 50222;
server_name    192.168.138.157;
    location ~/group([0-9])/M00 {
        root /data/fastdfs/data;
        ngx_fastdfs_module;
    }
}
```

修改 nginx.conf 文件后，将 Fast-Nginx 模块增加到 Nginx 中，命令如下。

```
[root@localhost nginx-1.15.2]# ./configure --add-module=/data/fastdfs-nginx-module/src
[root@localhost nginx-1.15.2]# make && make install
```

添加模块后，在 Nginx 目录下使用如下命令检查模块是否添加成功。

```
[root@localhost nginx-1.15.2]# /usr/local/nginx/sbin/nginx -V
```

添加成功后，检查命令返回应如下。

```
nginx version: nginx/1.15.2
built by gcc 4.8.5 20150623 (Red Hat 4.8.5-36) (GCC)
configure arguments: --add-module=/usr/local/mysoft/fastdfs-nginx-module/src
```

将 FastDFS 模块添加到 Nginx 后，需将 FastDFS 的部分配置信息放在/etc/fdfs 文件夹中，否则 Nginx 的 worker 进程无法正确启动，具体命令如下。

```
[root@localhost nginx-1.15.2]# cp /data/fastdfs-5.11/conf/http.conf    /etc/fdfs/
[root@localhost nginx-1.15.2]# cp /data/fastdfs-5.11/conf/mime.types   /etc/fdfs/
```

检查 /usr/local/nginx/conf/nginx.conf 文件配置是否正确，若正确，启动 Nginx 的命令如下。

```
## 启动
/usr/local/nginx/sbin/nginx
## 停止
/usr/local/nginx/sbin/nginx -s stop
## 重启
/usr/local/nginx/sbin/nginx -s reload
```

启动成功后 Nginx 进程应如下。

```
[root@localhost conf]# ps -ef|grep nginx
root       5220     1  0 09:51 ?        00:00:00 nginx: master process /usr/local/nginx/sbin/nginx
nobody     5221  5220  0 09:51 ?        00:00:00 nginx: worker process
root       5224  1138  0 09:51 pts/1    00:00:00 grep nginx
```

启动成功后，前台访问 Nginx 效果如图 11-10 所示。

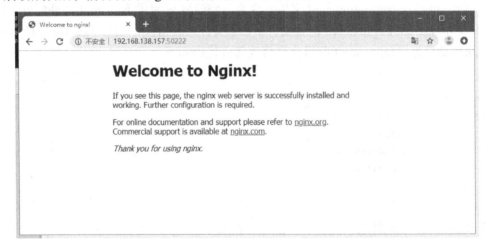

图 11-10

如果运行 Nginx 时，发生 /usr/local/nginx/sbin/nginx: error while loading shared libraries: libpcre.so.1: cannot open shared object file: No such file or directory 错误，就可能在安装 pcre 时，没有将相关的 lib 包放在 Linux 的 /lib 目录下（pcre 有时会安装在 /usr/local/lib/libpcre.so.1 和 /usr/local/lib/libpcre.so.1.2.8 目录下），此时需要执行下述两行代码以增加软连接。

```
[root@localhost nginx-1.15.2]# ln -s /usr/local/lib/libpcre.so.1 /lib64/
[root@localhost nginx-1.15.2]# ln -s /usr/local/lib/libpcre.so.1.2.8 /lib/libpcre.so.1
```

通过 Nginx 地址使用 HTTP 调取刚存储的图片，若部署成功，则直接在浏览器中展示图片，地址如下。

```
http://192.168.138.157:50222/group1/M00/00/00/wKiKnV21zb2AQiv7AABMLJJFkL8028.jpg
```

前台访问效果如图 11-11 所示。

图 11-11

如果 Nginx 整合 fastdfs-nginx-module 模块出现问题，可用以下命令删除 Nginx 并重新安装。

```
#删除
[root@localhost nginx-1.15.2] rm -rf /usr/local/nginx
#检查是否删干净
[root@localhost ~]# whereis nginx
nginx:
#检查返回 nginx:   则为完全删除 nginx 了
```

11.6 【实例】分布式微服务整合 FastDFS

11.6.1 实例背景

本实例将使用 boot_11_2 工程整合 FastDFS 的执行环境，并进行文件上传操作。创建相关项目及启动类，pom.xml 增加 FastDFS 依赖代码如下。

```xml
<!-- FastDFS 依赖 -->
<dependency>
    <groupId>com.github.tobato</groupId>
    <artifactId>fastdfs-client</artifactId>
    <version>1.26.5</version>
</dependency>
```

FastDFS 中关于 application.properties 资源配置文件代码如下。

```
fdfs:
  # 链接超时
  connect-timeout: 60
```

```
# 读取时间
so-timeout: 60
# 生成缩略图参数
thumb-image:
   width: 150
   height: 150
tracker-list: 192.168.138.157:50200
```

11.6.2 编写 FastDFS 核心配置类

编写 FastDFS 的核心配置类 DFSConfig.java 代码如下。

```
package com.zfx.config;
import com.github.tobato.fastdfs.FdfsClientConfig;
import org.springframework.context.annotation.Configuration;
import org.springframework.context.annotation.EnableMBeanExport;
import org.springframework.context.annotation.Import;
import org.springframework.jmx.support.RegistrationPolicy;
@Configuration
@Import(FdfsClientConfig.class)
@EnableMBeanExport(registration = RegistrationPolicy.IGNORE_EXISTING)
public class DFSConfig {}
```

@EnableMBeanExport(registration = RegistrationPolicy.IGNORE_EXISTING)注解是指通过@Import 将 JMX 相关的 Bean 定义加载到 IDC 容器中，当 Spring 容器已经存在的名称注册 MBean 时，应该忽略受影响的 MBean。若不编写该代码，则可能报错。

11.6.3 编写 FastDFS 工具类

编写 FastDFS 工具类以方便调用，工具类 FileDFSUtil.java 代码如下。

```
package com.zfx.util;
import com.github.tobato.fastdfs.domain.fdfs.StorePath;
import com.github.tobato.fastdfs.service.FastFileStorageClient;
import org.springframework.stereotype.Component;
import org.springframework.util.StringUtils;
import org.springframework.web.multipart.MultipartFile;
import javax.annotation.Resource;
@Component
public class FileDFSUtil{
    @Resource
    private FastFileStorageClient storageClient ;
    /**上传文件*/
    public String upload(MultipartFile multipartFile) throws Exception{
        String originalFilename = multipartFile.getOriginalFilename().
                            substring(multipartFile.getOriginalFilename().
                            lastIndexOf(".") + 1);
```

```java
            StorePath storePath = this.storageClient.uploadImageAndCrtThumbImage(
                            multipartFile.getInputStream(),
                            multipartFile.getSize(),originalFilename , null);
            return storePath.getFullPath();
    }

    /**删除文件*/
    public void deleteFile(String fileUrl) {
        if (StringUtils.isEmpty(fileUrl)) {return;}
        try {
            StorePath storePath = StorePath.parseFromUrl(fileUrl);
            storageClient.deleteFile(storePath.getGroup(), storePath.getPath());
        } catch (Exception e) {
            e.printStackTrace();
        }}}
```

11.6.4 编写测试接口

为方便测试，编写 FilController.java 接口类，FilController.java 接口类提供文件上传与文件删除接口，其代码如下。

```java
package com.zfx.controller;
import org.springframework.util.StringUtils;
import org.springframework.web.bind.annotation.PathVariable;
import org.springframework.web.bind.annotation.RequestMapping;
import org.springframework.web.bind.annotation.RequestMethod;
import org.springframework.web.bind.annotation.RequestParam;
import org.springframework.web.bind.annotation.RestController;
import org.springframework.web.multipart.MultipartFile;
import javax.annotation.Resource;
@RestController
public class FileController {
    @Resource
    private FileDFSUtil fileDFSUtil ;
    /**文件上传*/
    @RequestMapping(value = "/uploadFile",headers="content-type=multipart/form-data", method = RequestMethod.POST)
    public String uploadFile (@RequestParam("file") MultipartFile file){
        String result ;
        try{
            String path = fileDFSUtil.upload(file) ;
            if (!StringUtils.isEmpty(path)){
                result = path ;
            } else { result = "上传失败" ;}
        } catch (Exception e){
```

```
                e.printStackTrace();
                result = "服务异常";
            }
        return result;
    }
    /**文件删除*/
    @RequestMapping(value = "/deleteByPath/{filePathName}", method = RequestMethod.GET)
    public String deleteByPath (@PathVariable String filePathName){
        fileDFSUtil.deleteFile(filePathName);
        return "删除成功";
    }}
```

文件查看与下载直接依靠 HTTP 访问 Nginx 下路径即可。在实际项目中，相关的 filePath 与 name 等信息应存储在 MySQL 等关系型数据库中，进行持久化存储与增删改等操作。

11.6.5 当前项目结构

当前项目结构如图 11-12 所示。

11.6.6 运行结果

启动 boot_11_2 项目，打开 Swagger 页面，通过 uploadFile 接口上传静态资源文件，如图 11-13 所示。

单击 Try it out 按钮将任意文件上传至 FastDFS，接口返回结果如图 11-14 所示。

图 11-12

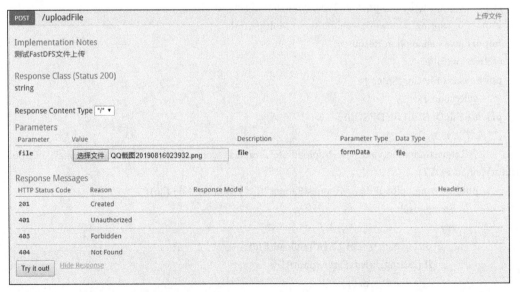

图 11-13

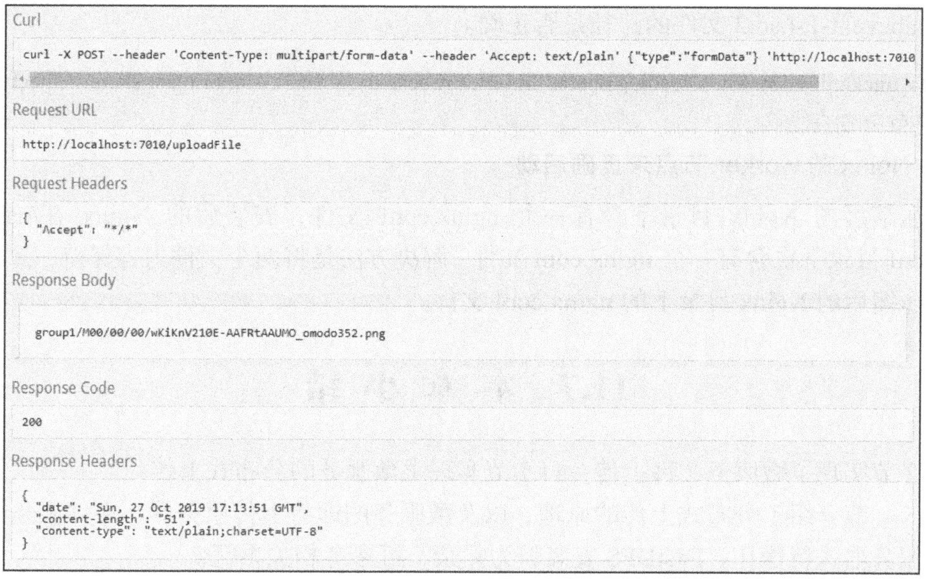

```
Curl

curl -X POST --header 'Content-Type: multipart/form-data' --header 'Accept: text/plain' {"type":"formData"} 'http://localhost:7010

Request URL

http://localhost:7010/uploadFile

Request Headers

{
  "Accept": "*/*"
}

Response Body

group1/M00/00/00/wKiKnV210E-AAFRtAAUMO_omodo352.png

Response Code

200

Response Headers

{
  "date": "Sun, 27 Oct 2019 17:13:51 GMT",
  "content-length": "51",
  "content-type": "text/plain;charset=UTF-8"
}
```

图 11-14

在 Response Body 文本框中可以看到该文件存储的位置，通过 HTTP 请求该地址即可返回之前存储的静态资源文件。

11.6.7 实例易错点

1. FastDFS 执行异常

FastDFS 执行异常日志如下，在 Linux 系统执行过程中，一旦含有 Permission denied 关键字的错误，大多是由权限有误引起的。

```
[2019-03-05 22:32:20] ERROR - file: ../storage/trunk_mgr/trunk_shared.c, line: 177, "Permission denied" can't be accessed, error info: /home/kenick/fastDFS_home_v5.05
2019/03/05 22:32:20 [alert] 8061#0: worker process 8062 exited with fatal code 2 and cannot be respawned
    ngx_http_fastdfs_process_init pid=8147
```

解决方法是赋予 Fast 相应权限或在 nginx.conf 中增加权限设置如下。

```
user   root
```

2. FastDFS 端口号冲突引起的异常

FastDFS 端口号冲突引起的异常日志如下，此时需要检测打开端口，或关闭防火墙，或检查 FastDFS 是否正常运行。

```
ERROR - file: connection_pool.c, line: 130, connect to 192.168.0.8:23000 fail, errno: 113, error info: No route to host
```

3. FastDFS 缺失文件或依赖包

出现 No such 关键字的异常如下，大多由 FastDFS 缺失相关文件或依赖包引起，此时应检查 libevent-1.4.so.2 文件的路径是否正确。

> /fdfs_trackerd: error while loading shared libraries: libevent-1.4.so.2: cannot open shared object file: No such file or directory

4．Nginx 的 worker 节点未正确启动

解压缩后的 Nginx 目录下含有一个 nginx.conf 文件，安装后的 Nginx 在/usr/local/nginx/conf 目录下也含有一个 nginx.conf 文件。解决方法是将两个文件内容保持一致，即需修改解压缩后的 Nginx 目录下的 nginx.conf 文件。

11.7 本章小结

11.2 节实现了微服务文件上传，11.7 节实现了微服务的分布式上传。

11.5.4 节介绍了分布式上传的原理，以及微服务的即将上传服务器、文件服务器分割成两个服务器进行操作。FastDFS 安装较为复杂，可参考相关书籍。

11.8 习　　题

1．创建 Spring Boot 工程，使用 Spring Boot 整合文件上传，在前台使用 JSP 页面上传文件后，微服务的接口将上传的文件存储到指定地点。

2．搭建 FastDFS 分布式文件系统，搭建后使用命令可具备文件上传/下载等相关能力。

3．创建 Spring Boot 工程，使用 Spring Boot 整合 FastDFS 分布式文件系统，可具备文件上传/下载等相关能力。

4．整理并分析分布式与集群之间的相同点和不同点。整理并分析分布式任务调度 Quartz、分布式文件系统 FastDFS、分布式注册中心 Consul、Spring Cloud 的相同点与不同点。

第 12 章　扩展与部署

微服务分布式架构的过程与本书的章节顺序基本相同，即先构建一个简易的微服务，然后将该微服务注册到注册中心上，再分别在该微服务上集成通信、断路器、负载均衡等基础功能，在微服务上进行持久化、缓存等操作，最后通过任务调度、异步线程、分布式通信、文件上传/下载等完善程序。

本章对微服务分布式架构进行扩展与总结。

12.1　微服务分布式架构相关方案总结

微服务群、注册中心集群、MySQL 集群（分库分表）、分布式事务、MQ 消息中间件及集群、文件存储管理器集群、任务调度集群（定时任务集群）、Java Web 客户端负载均衡/服务端负载均衡集群、Redis 缓存集群、搜索引擎集群、响应式 Webflux、Kubernetes+Docker 监控与部署都属于分布式的架构技术，目的是使用架构的方式将所需压力转移到其他服务器上或多个服务器共同承担性能压力。

12.1.1　解决方案与目标

微服务群通过多台服务器共同协作的方式，解决了单个 Java Web 应用程序在某个服务器上所承担的性能压力，将各 Service 分到了不同的服务器中，减少每台服务器的所需线程，减小单台服务器的性能压力，以此提高应用程序的总体性能，并增加了更加完善便捷扩展的能力。相关技术有 Spring Cloud+Spring Boot 微服务、Dubbo 服务等，微服务群和服务群都依靠注册中心进行交互。

注册中心集群通过多台服务器负载的方式，解决了单个注册中心的性能瓶颈问题，原本单个注册中心需长轮询多个微服务并被其调用提供服务地址等相关信息。若微服务注册数目与调用次数过多，单台注册中心则难以消化全部进程，所以在多台服务器同时负载进行处理时，分流微服务的请求，提高了总体注册中心的性能和注册中心宕机后的高可用性。相关技术有 Zookeeper、Eureka、Consul，通常各注册中心自身都有集群功能。

MySQL 集群通过多台服务器负载的方式将单个 MySQL 进行分库分表处理，提高了 MySQL 单表存储上限，并且在 MySQL 集群搭建后，请求会根据 MyCat 分流到不同的 MySQL 中，减少单台 MySQL 的被请求线程和请求数目，提高了总体关系型数据库的存储空间及性能。为 MySQL 制作集群的方案和相关技术有 MyCat、MHA、Haproxy+Keepalived 等。

MQ 一系列消息中间件将消费者/生产者模式从代码中转到了单独服务器中，通过转移压力的方式，将部分所需要停留的数据与请求额外放置在其他服务器中，达到提高 Java Web 应用程序性能的目的。相关技术有 ActiveMQ、RabbitMQ、Kafka、RocketMQ 等，通常各消息队列自身都有集群功能。

MQ 中间件集群通过多台服务器负载的方式，减小每台服务器存储与被请求的性能压力，提高总体消息对垒的性能，解决了单台 MQ 类消息中间件的性能上限与高可用性问题。

文件存储管理将文件上传后的存储从本服务器上解放到了其他服务器上，通过转移压力提高应用程序性能及存储空间的能力。相关技术有 FastDFS、HDFS、TFS、GFS、NFS 和 MogileFS。文件存储集群通过多台服务器负载的方式，提高了文件存储的上限（只需要加卷改配置即可，不需要修改代码）。

任务调度集群通过多台服务器协作的方式，提高了单任务调度的高可用性，防止宕机。相关技术有 Opencron、LTS、XXL-Job、Elastic-Job、Uncode-Schedule、Antares、SkySchedule、timerTask、Quartz 等。

Java Web 客户端负载均衡/服务端负载均衡通过多台服务器负载均衡的能力，将请求进行分流，减少每台 Java Web 应用程序的被请求线程和数目，提高整体应用程序的并发性能。相关技术有 LVS+Keepalived+Nginx（整体或单独使用都可）、Spring Cloud Ribbon 等。

Redis 缓存集群、搜索引擎集群通过算法使用多台服务器将数据进行冗余存储等方式，将多台服务器进行负载，减少单台服务器被请求和线程数目，并且多台服务器的冗余存储会防止宕机。相关技术有 Redis、MongoDB、EVCache、Memcached、Elasticsearch、Solr 等，以上都可配置成集群模式。

12.1.2 分布式部分技术细节扩展

分布式架构大多通过多台服务器协作、负载、缓存、业务拆分（微服务）、数据拆分（分库分表）等方式，分摊总体压力，达到提高总体服务性能的目的。

分布式还有前后端分离、数据库读写分离与动静分离、主从分离、应用层与数据层分离、CDN 加速、异步架构、响应式架构、冗余化管理、灰度发布、单点登录、页面静态化、微服务网关等相关知识，可参考其他相关资料。

12.1.3 动静分离

动静分离是指将动态代码与静态文件分离，用 Nginx、Apache 等相关服务器管理图片、音频等相关内容，JavaScript 只调用静态资源的 URL 地址，不直接将资源嵌入前端页面中。通常静态分离的图片服务器集群会使用独立的二级域名。

12.1.4 前后端分离

最初分布式 Java Web 项目前后端不分离，前端人员将写好的 HTML 页面发送给后端人员，由后端人员转换成 JSP 文件，并对其进行修改，但日后维护页面效果时，双方代码差异太大，并且沟通成本过高，由此引出前后端分离的架构模式。

前后端分离的开发模式是指前端以 HTML5 为主，使用 VUE.js、React.js、Angular.js 等前端框架，前端人员可以在观察后端人员 Swagger 的接口文档后，调用后端接口直接使用类似$符号的方式获得相关数据，使用简捷，与 JSP 引擎一样减少了大量的 JavaScript 代码，并减少了沟通成本，在后期维护上不需要前后台互相参与。在部署上只需要用不同的前端框架打包，使用 Nginx 进行转发即可。

当前移动端架构方式较多。例如，只使用 HTML5 构建一套前端代码，分别使用 Android 与 iOS 的 WebView 控件对其集成 HTML5，即可只写一套代码在不同系统中运行。HTML5 内可以使用 Ionic、JQuery Mobile、Bootstrap 进行混合开发，或使用 Unity、Cocos、Egret 等游戏引擎进行开发。

12.1.5　数据库读写分离与主从分离

读写分离是指将数据库分成了主从库，即一个主库用于输入数据，多个从库完成读取数据的操作，主从库之间通过某种机制进行数据同步，是一种常见的数据库架构。读写分离的目的是解决数据库的读取瓶颈。

12.1.6　应用层与数据层分离

应用层与数据层分离是指将整个应用程序分为应用服务器、文件服务器、数据库服务器、搜索引擎服务器、缓存服务器等，将数据存储压力分散。以上服务器对硬件资源的要求各不相同，应用服务器由于要处理大量的业务逻辑和线程，所以需要更好的 CPU；数据库服务器和搜索引擎服务器需要快速磁盘检索和数据存储，所以需要更好的内存；文件存储服务器则需要容量更大的硬盘。

应用层与数据层分离后，除将压力分散外，最大的优势是可以针对不同的服务器配置不同的硬件资源，将资源最大化利用。

12.1.7　CDN 加速

CDN 加速部署在网络服务提供商的机房，使用户在请求网络服务时，可以从距离自己最近的网络提供商的机房获取数据，使用 CDN 加速的目的是尽可能将数据传输距离缩短，加快数据的返回速度。通常，各运营商都出售 CDN 加速服务。

除 CDN 加速外，一些面向全国甚至全球的 Web、APP 需要类似城域网的架构，根据当前登录 IP 区分城市/省会请求到不同的服务器上，缩短数据传输的距离。

12.1.8　异步架构

此处的异步是指以消息队列为主的异步架构模式，异步架构是典型的生产者/消费者设计模式，两者不存在直接调用。生产者的微服务将数据缓存在其他服务器后，消费者的微服务将依据执行条件消费数据，减轻数据/线程在本机的资源消耗，提高系统可用性，加快网站的响应速度，消除并发访问高峰。

12.1.9　响应式编程

响应式编程特点如下。

（1）灵敏：任何情况下系统都尽可能用最快速度响应。

（2）可恢复：任何情况下系统出现任何问题都有相应的容错机制。

（3）可伸缩：请求多时使用扩展算法和软硬件的方式增加自身性能，请求少时释放资源。

（4）消息驱动：响应式编程存在异步机制，通过消息连接各事件并相互协作。

基于上述特点，响应式编程提出了各种模型满足响应式编程的理念，其中最出名的有 Reactor 和 RxJava，Spring 基于 Reactor 和 RxJava 模型构建了 WebFlux 响应式编程框架。

Spring WebFlux 是一种新型框架，与微服务分布式的编程过程略有区别。Spring WebFlux 提供了一种声明式流定义语言，用于在更高的抽象级别上创作流。它允许将其集成到广泛的应用程序中，而不需要任何更改流编程模型，包括 SpringMVC、JSF，以及 PortletWeb 应用程序。

Spring WebFlux 为以下问题提供了解决方案。

（1）可视化流程非常困难。

（2）应用程序有很多代码访问 HTTP 会话。

（3）正确的浏览器返回按钮无法实现。

（4）浏览器、服务器与后退按钮的使用不同步。

（5）多个浏览器选项卡导致 HTTP 会话数据的并发。

例如，在签入航班、申请贷款、购物车结账的表单提交中，可以使用 Spring WebFlux，通常这样的表单场景有以下特性。

（1）明确的起点和终点。

（2）用户必须按特定顺序通过一组屏幕。

（3）这些更改直到最后一步才能完成。

（4）一旦完成就不可以意外地重复事务。

12.1.10 冗余化管理

冗余化管理是指即使某个服务或单元被调用的次数很少，也要配置成高可用集群的模式，将意外扼杀在萌芽中。

在冗余化管理的基础上，可以增加交叉管理的方式进行部署。例如，某几个单元或服务都很少被调用，可以把这几个单元交叉部署，只要不出现所有服务器全部宕机的情况，即可正常运行整体服务。

12.1.11 灰度发布

灰度发布（也称金丝雀发布）是指在黑与白之间，能够平滑过渡的一种发布方式。在其上可以进行 A/B 测试，即让一部分用户继续用产品特性 A，一部分用户开始用产品特性 B，如果用户对 B 没有反对意见，那么逐步扩大范围，把所有用户都迁移到 B 上。灰度发布可以保证整体系统的稳定，在初始灰度时就可以发现、调整问题，以保证其影响度。灰度发布从开始到结束的一段时间，称为灰度期。

12.1.12 页面静态化

页面静态化是指通过 Freemaker、Thymeleaf 等相关静态化模板引擎，将 JSP、ASP、

PHP 等动态页面转换为 HTML 页面，并将 HTML 页面使用 FTP 等协议直接存储在某个可映射路径下，待下次读取该页面时可直接读取 HTML 页面，节省 JSP、ASP、PHP 等动态页面的渲染时间、接口调用时间。

页面静态化通常用在门户网、小说网、官网、新型商品展示页等不经常修改但频繁调用的页面。

12.1.13　服务端主动推送

早期使用 Ajax 向服务器轮询消息，使浏览器尽可能第一时间获取服务端接口的信息，若同时在线用户过多，可能会大大增加服务端的压力。目前，服务端主动推送技术分为以下两种。

（1）客户端向服务端发送请求，服务端会抓住该请求并不进行释放或返回操作，等有数据更新时才返回客户端，客户端接收消息后，再向服务端发送请求，周而复始。

（2）使用持久化双向监听技术 WebSocket，WebSocket 原理即在页面处放置一个 Socket 端口，服务器可直接将数据传输到该 Socket 端口处，该协议是 HTML5 发布后的一个新协议。

12.2　微服务扩展

12.2.1　微服务整合日志

Spring Boot 集成了 Log4j、Log4j2、Logback、Logging 作为日志管理工具。无论使用哪种日志工具，Spring Boot 微服务都已经将其整合了相关的控制台输入与文件输出的默认配置。若希望更改其配置，则需要增加不同日志工具的 properteis 或 XML 等配置文件。

12.2.2　微服务整合单元测试

Spring Boot 集成了 Junit 单元测试，也可通过如下依赖增加 Mockito 工具，以增加 Mock 测试能力。

```
<dependency>
    <groupId>org.mockito</groupId>
    <artifactId>mockito-all</artifactId>
    <version>1.8.5</version>
    <scope>test</scope>
</dependency>
```

在程序的 Service 过于复杂，或不能调用 Service 接口时，可以使用 Mockito 工具的 Mock 对象对其进行测试。Mock 对象在调试期间作为真实对象的替代品，Mock 测试是在测试过程中对那些不容易构建的对象用一个虚拟对象来代替测试的方法。

12.2.3　微服务整合全局异常

Spring Boot 集成了全局异常注解@ControllerAdvice，将控制器的全局配置放置在同一个位置，所有使用@Controller 注解修饰的类，都可以使用@ExceptionHandler、@InitBinder、@ModelAttribute 注解对@Controller 接口进行全局异常控制。一旦@Controller 接口发生异常，都可以跳转到相应的 Exception.class 中。

12.2.4　微服务整合 JSR-303 验证机制

Spring Boot 集成了 JSR-303 验证规范，可协助程序员对 POJO/Java Bean 进行数据认证的编程，伪代码如下。

```
@NotNull(message ="name 不能为空")
private String name;

@DateTimeFormat(pattern = "yyyy-MM-dd")// 日期转化格式
private Date date;

@NotNull
@DecimalMin(value = "0.1")// 最小值为 0.1
@DecimalMax(value = "10000.00")// 最大值为 10000
private Double doubleValue = null;

@NotNull
@Min(value = 1,message="最小值为 1")
@Max(value=88,message="最大值为 88")
private Integer integerValue;

@Range(min = 1,max=888,message="范围为 1 至 888")
private Long rangeValue;

@Email(message="邮箱格式认证错误")
private String emailValue;

@Size(min=20,max=30,message="字符串长度要求 20 到 30 之间")
Private String sizeValue;
```

12.2.5　微服务整合国际化

Spring Boot 集成了 SpringMVC 的 MessageSource 接口体系，实现国际化消息源。

12.2.6　微服务整合安全与认证

Spring Boot 集成了 Spring Security 作为安全与认证工具，Spring Security 是安全类框架，利用 Spring 的 IOC 与 AOP 实现认证与授权的相关功能。认证即明确用户可以访问、登录、使用该系统，授权即明确该用户在当前系统中所包含的功能。

12.2.7　微服务整合 WebSocket 协议

Spring Boot 集成了 WebSocket 协议，WebScoket 协议是在 HTML5 发行后产生的，基

于 TCP 的一种新型网络协议。WebSocket 协议实现了阅览器与服务器双面通信的能力,即允许服务端主动发送消息给客户端。

WebSocket 协议将原本 TCP/IP 的三次握手改成了长连接的形式,让客户端和服务端在初步握手后,不需要再次握手即可来回发送信息,直到(手动/自动/意外)断开连接。部分没有实现 WebSocket 协议的协议还需要通过 STOMP 完成对 WebSocket 协议的兼容。

12.2.8 微服务整合 HTTPS

Spring Boot 集成了 HTTPS,在实际运行中需要先购买 HTTPS 的 CA 证书,在服务端执行证书命令后,在 Spring Boot 微服务工程中的 application.properties 资源配置文件里修改关于 HTTPS 的配置代码如下。

```
#https
server.port: 8443
server.ssl.key-store: classpath:message.p10
server.ssl.key-store-password: mypassword
server.ssl.keyStoreType: PKCS12
server.ssl.keyAlias: message.p10
```

12.2.9 微服务整合批处理

Spring Boot 集成了 Spring Batch,用来处理相对大型的数据。Spring Batch 主要用来读取大量数据,然后经处理后输出成指定的形式。

Spring Batch 只需要注册成 Spring Bean 即可使用,并在微服务启动类上使用 @EnableBatchProcessing 注解即可支持批处理功能。

12.2.10 微服务整合 lombok

Spring Boot 集成了 lombok,lombok 是指增加一些"处理程序",可以使 Java 变得简捷。

在使用 lombok 注解前,需要在 Eclipse 或 Idea 等编程工具上安装相应的 lombok 插件。常用的 lombok 注解有@Data、@Setter、@Getter、@EqualsAndHashCode、@Builder 等,可以帮助程序减少 Setter、Getter 等相关代码。

12.2.11 微服务整合异步消息驱动

Spring Boot 集成了 Spring Integration,提供了异步消息驱动的能力,主要用来解决程序之间交互的问题,其原理类似于 9.3 节中的使用 Redis 实现消息通信。

Spring Integration 主要由 Message、Channel、Message EndPoint、ChannelInterceptor 四部分组成。Message 是传递的数据,Channel 是消息发送者与消息接收者之间交流的管道,Message EndPoint 是处理消息的组件,ChannelInterceptor 是 Channel 的拦截器。

12.2.12 分布式链路监控

分布式链路监控的特点包括低性能损耗、尽可能不埋点、可伸缩性、可视化、迅速反馈、可持续性及高可用性。Spring Cloud 通过 Spring Cloud Sleuth 作为分布式链路监控的解决方案，Spring Cloud Sleuth 将自动采集系统之间的交互信息，用户可以通过日志文件获取链路数据，也可以将数据发送给远程服务进行统一收集。

12.2.13 分布式单点登录

分布式单点登录 SSO 包括多种解决方案，如下所述。

（1）共享 Session 方案：最初的分布式解决方案，其含义是将用户的 Session 信息存储到 Redis 之类的缓存中，以用户的 Key 值查看用户是否已经登录，若用户已经登录，则不生成新的 Session，将原有的 Session 返回给用户。

（2）Session 与网关结合方案：类似共享 Session 方案，用户在网关处进行登录，登录后查看 Redis 等缓存中是否含有用户 Session，若含有则后端从缓存中获取相关 Session。

（3）Token+Session+网关结合方案：客户端持有 token 通过网关与服务端发生请求，若校验 token 通过，则通过 token 到 Redis 中获取相关 Session 返回给用户。

（4）非对称口令+Token+Session+网关结合方案：此方案分为两步，执行前客户端需持有 RSA 非对称口令的公钥，先用公钥加密后的字符串获得 token，再请求服务端获得相关 token 并授权登录。

（5）Cookie 缓存+网关结合方案：通过浏览器将用户信息存储在 Cookie 中，然后通过网关或服务端解析 Cookie，从而获得用户相关信息，但此方案较不安全。

12.3 【实例】分布式网关的初步测试

12.3.1 实例背景

Spring Cloud 在较早期版本中推荐使用 Spring Cloud Zuul 作为分布式网关，后期更换为 Spring Cloud Gateway。

分布式网关是为了保证动态路由、监控、弹性、服务、安全、拦截器、限流等功能做出的方案。在前端进行访问时，网关将基于 JVM 路由器为服务端提供可配置的对外 URL 到服务的映射关系。网关提供了丰富的 API 全程托管，降低管理成本和安全风险，包括协议适配、协议转发、安全策略、防恶意刷新、防恶意消耗流量等安全措施。Spring Cloud Gateway 通过 Gateway 路由、Gateway 断言工厂、Gateway 过滤器等组件以具备网关概念上的相应功能。

本实例通过创建 gateway-test-01 工程与 gateway-test-02 工程，分别用函数注册与资源配置文件的方式，将 localhost:8080/baidu 地址转换到 https://www.baidu.com 地址。通过初步测试介绍分布式网关。

12.3.2 使用资源配置文件的方式配置分布式网关

gateway-test-01 工程的启动类 ApplicationMain.java 与正常启动类相同，gateway-test-01 工程的 application.yml 资源配置文件代码如下。

```yaml
server:
  port: 8080
spring:
  application:
    name: api-gateway
  cloud:
    gateway:
      routes:
        - id: gateway-service
          uri: https://www.baidu.com
          predicates:
            - Path=/baidu
```

gateway-test-01 工程的目录结构如图 12-1 所示。

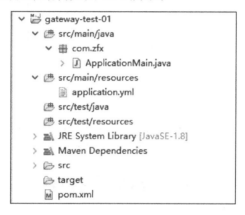

图 12-1

12.3.3 使用注册 Bean 的方式配置分布式网关

gateway-test-02 工程只需要增加 GateWayConfig.java 配置类即可，其代码如下。

```java
package com.zfx.config;
import org.springframework.cloud.gateway.route.RouteLocator;
import org.springframework.cloud.gateway.route.builder.RouteLocatorBuilder;
import org.springframework.context.annotation.Bean;
import org.springframework.context.annotation.Configuration;
@Configuration
public class GateWayConfig {
    @Bean
    public RouteLocator routeLocator(RouteLocatorBuilder builder) {
        return builder.routes()
                .route(r ->
                        r.path("/baidu")
                                .filters(f -> f.stripPrefix(1))
                                .uri("http://www.baidu.com")
                ).build();
```

        }
    }

gateway-test-02 工程的目录结构如图 12-2 所示。

图 12-2

12.3.4 运行结果

启动 gateway-test-01 工程或 gateway-test-02 工程后，通过浏览器输入 localhost:8080/baidu 地址即可跳转到 https://www.baidu.com/地址。

12.4 微服务打包

12.4.1 Jar 包

Spring Boot 微服务工程需要打 Jar 包时有以下两种选择方案。

（1）在 Eclipse 等 IDE 工具中直接使用 export 打成 Jar 包，此种打包方式可能会出现静态资源文件路径错误。

（2）使用 MVN 命令行，进入相关目录后，使用 maven package 命令进行打包。

Spring Boot 微服务在打包后，将 Jar 包传输到服务器，使用 java -jar xx.jar 命令可直接运行微服务。如果需要将 Jar 包注册为 Linux 服务，需要更改关于 pom.xml 文件的依赖配置代码如下，再使用 maven package 命令进行打包，最后使用 Linux 系统的 init.d 或 systemd 注册服务。

```
<build>
    <plugins>
        <plugin>
            <groupId>org.springframework.boot</groupId>
            <artifatId>spring-boot-maven-plugin</artifatId>
            <configuration>
                <executable>true</executable>
            </configuration>
        </plugin>
```

```
    </plugins>
  </build>
```

12.4.2 War 包

Spring Boot 微服务工程在打 War 包前需要更改 pom.xml 文件代码如下。

```
<packaging>war</packaging>
<dependencies>
  <dependency>
    <groupId>org.springframework.boot</groupId>
    <artifactId>spring-boot-starter-tomcat</artifactId>
    <scope>provided</scope>
  </dependency>
</dependencies>
```

更改 pom.xml 文件代码后,增加 SpringBootServletInitializer 的继承实现类代码如下。

```
Import org.springframework.boot.builder.SpringApplicationBuilder;
Import org.springframework.boot.context.web.SpringBootServletInitializer;

public class ServletInitializer extends SpringBootServletInitializer{
    @Override
    Protected SpringApplicationBuilder configure(SpringApplicationBuilder application){
        return application.sources(SpringApplicationMain.class);
    }
}
```

此后使用 maven package 命令进行打包,即可获得 War 包。

12.5 本章小结

本章拓展了分布式架构的相关方案,总结了本书前几章未涉及的 Spring Boot 与 Spring Cloud 框架扩展内容。

12.6 习　　题

1. 创建 Spring Boot 工程,初步整合分布式网关 Spring Cloud Gateway。
2. 将微服务打为 Jar 包并在 Linux 服务器上运行。
3. 将微服务打为 War 包并在 Linux 服务器上运行。

参 考 文 献

[1] 汪云飞. Java EE 开发的颠覆者 Spring Boot 实战[M]. 北京：电子工业出版社，2016.

[2] 翟永超. Spring Cloud 微服务实战[M]. 北京：电子工业出版社 2017.

[3] 杨恩雄. 疯狂 Spring Cloud 微服务架构实战[M]. 北京：电子工业出版社 2018.

[4] 许进，叶志远，钟尊发，等. 重新定义 Spring Cloud 实战[M]. 北京：机械工业出版社，2018.